HOPF SPACES

NORTH-HOLLAND
MATHEMATICS STUDIES 22

Notas de Matemática (59)
Editor: Leopoldo Nachbin

Universidade Federal do Rio de Janeiro
and University of Rochester

Hopf Spaces

ALEXANDER ZABRODSKY
Associate Professor, Hebrew University, Jerusalem, Israel

1976

NORTH-HOLLAND PUBLISHING COMPANY - AMSTERDAM • NEW YORK • OXFORD

North-Holland ISBN: 0 7204 0553 X

PUBLISHERS:
NORTH-HOLLAND PUBLISHING COMPANY
AMSTERDAM • NEW YORK • OXFORD

DISTRIBUTORS FOR THE U.S.A. AND CANADA:
AMERICAN ELSEVIER PUBLISHING COMPANY, INC.
52 VANDERBILT AVENUE, NEW YORK, N.Y. 10017

Library of Congress Cataloging in Publication Data

Zabrodsky, Alexander.
Hopf spaces.

(Notas de matemática ; 59) (North-Holland mathematics studies ; 22)
Bibliography: p.
Includes index.
1. H-spaces. I. Title. II. Series.
QA612.77.Z3 512'.55 76-41394
ISBN 0-7204-0553-X

PRINTED IN THE NETHERLANDS

Table of Contents

Possibly more than any other field in mathematics algebraic topology contains a vast amount of relatively simple facts, clusters of small theorems and intuitive observations. Naturally, these include many folklore-type theorems which do not appear in the literature. (It is quite likely that as many theorems in algebraic topology have appeared verbally in universities' common rooms as have appeared in print in the professional literature). It is therefore essential to bring some of these fundamentals to light in print from time to time.

The subject of H-spaces within the field of algebraic topology is no exception. In the last decade some outstnading progress has been made on the subject, affecting related fields in homotopy theory such as the theory of cohomology operations, classifying spaces, infinite loop spaces and localization theory.

These notes try to describe some of these new developments. We make no attempt to encompass all areas of progress. Instead, we concentrate only on three subjects: the structure of the cohomology of H-spaces, the mod p theory of H-spaces and applications of the relatively new BP theory in the study of H-spaces.

An attempt has been made to bring a substantial part of these notes to within the grasp of graduate students and algebraic topologists who do not specialize in this particular subject.

The first two chapters of these notes cover the fundamental concepts and hence, are essential for the understanding of the last three. These last, however, are fairly independent.

The only systematic treatment of the subject of H-spaces in the literature is Stasheff's "H-spaces from the homotopy point of view" ($[\text{Stasheff}]_5$). This was written during a period of rapid development in the field and some of the newer results were still unpolished. There is naturally some overlap between this work and Stasheff's notes, notably in the first two chapters of this book. The principal distinction between the two volumes is that the specific aspects of H-space theory treated in detail are clearly different: We do not consider here subjects such as projective planes, classifying spaces, homology operations and infinite loop spaces.

Some relationship can be found between Chapter IV of these notes and [Hilton, Mislin, Roitberg].

While preparing these notes the author was partly supported by a grant from the ETH of Zurich and by the British Science Research Council to whom I would like to express my gratitude. I also would like to thank the members of the Forschungsinstitut fur Mathematik and the Mathematics Department of the ETH and the members of the Mathematical Institute of Oxford University for their patience in discussing with me these notes in their various stages of production.

Alexander Zabrodsky

The Institute of Mathematics
The Hebrew University
JERUSALEM

Chapter 0

Notations, Conventions and Preliminary Observations

0.1. Spaces and maps

All spaces in these notes are assumed to be simply connected and of the homotopy type of CW complexes of finite type. All have a non-singular base point. We use the notation $*$ for base points of all spaces (thus considering a base point as a map $*$ from the singleton $*$ to X with image $* \in X$).

Subspaces are always assumed to be NDRs and one can always apply the homotopy extension property.

All maps are pointed (i.e.: base point preserving). All subspaces contain the base point. Composition of maps are denoted by juxtaposition: $f \circ g = fg$.

We use the customary notations

$$X \vee X = \{x,y \in X \times X \mid * \in \{x,y\}\}, \qquad X \wedge X = X \times X/X \vee X$$

We denote by Λ the identification map $\Lambda\colon X \times X \longrightarrow X \wedge X$.

We use as a standard notation Δ for the diagonal map: $\Delta = \Delta_X\colon X \to X \times X$, $\Delta(x) = x,x$ and Σ for the suspension: $\Sigma X = S^1 \wedge X$.

0.2. Homotopies

All homotopies are pointed, i.e.: $F\colon X \times I \to X'$ always satisfies $F(*,t) = *$, $0 \leq t \leq 1$.

If $f: X_0 \to X_1$, $g_0, g_1: X_1 \to X_2$, $h: X_2 \to X_3$ then a homotopy $F: X_1 \times I \to X_2$ of g_0 and g_1 ($F: g_0 \sim g_1$, $F(x,\varepsilon) = g_\varepsilon(x)$, $\varepsilon = 0,1$) is said to be *relative to* f *(abr.* rel f) if $F(f \times 1) = g_0 f p_1 = g_1 f p_1$ i.e. $F(f(x),t) = g_0 f(x) = g_1 f(x)$. F is said to be *corelative to* h *(corel* h*)* if $hF = hg_0 p_1 = hg_1 p_1$, i.e.: $hF(x,t) = hg_0(x) = hg_1(x)$.

If $f: X_0 \subset X_1$ is an inclusion of a subspace, we sometimes write $g_0 \sim g_1$ rel X_0 instead of rel f.

0.3. Categories and adjoint maps

We occasionally use categorical notations, but our category theory never goes beyond the phase of a notational system.

We work in the category of pointed (homotopy types of) CW complexes and continuous maps and not in the homotopy category (where homotopy classes of maps are the morphisms). We avoid the homotopy category because the homotopies themselves are being lost in the homotopy category. These homotopies are needed to obtain invariants and obstructions throughout these notes.

Nevertheless, quite often we identify ambiguously maps with their homotopy classes and thus mark as equal homotopic maps. (This is done more often in the last three chapters.) Commutative diagrams always commute only up to homotopy.

As usual we denote by $[X,Y]$ the set of homotopy classes of (pointed) maps $X \to Y$. $[f]$ denotes the homotopy class of f.

If $h: X' \to X$, $g: Y \to Y'$ we write $h^*: [X,Y] \to [X',Y]$, $g_*: [X,Y] \to [X,Y']$ for the functions induced by h and g (corresponding to the categorical notations of $[h,Y]$ and $[X,g]$ respectively).

Given X_1, X_2, Y (X_2 locally compact) there exists a natural equivalence $[X_1 \wedge X_2, Y] \longrightarrow [X_1, Y^{X_2}]$ where Y^{X_2} is the space of pointed maps $X_2 \to Y$. The equivalence is given by assigning to $[f]$, $f\colon X_1 \wedge X_2 \longrightarrow Y$ the class of $(f)_\#\colon X_1 \to Y^{X_2}$, $(f)_\#(x)[x'] = f(x,x')$. We denote by $(\ )_\#$ the inverse assignment as well: If $h\colon X_1 \longrightarrow Y^{X_2}$ then $h_\#\colon X_1 \wedge X_2 \longrightarrow Y$ is given by $h_\#(x_1,x_2) = h(x_1)[x_2]$. We refer to $f_\#, h_\#$ as *the adjoints* of f and h (thus omitting the left and right distinction of adjoints).

We shall only use these notions for the cases $X_2 = I$ or $X_2 = S^1$ (where we have $_\#\colon [\Sigma X,Y] \longrightarrow [X,\Omega Y]$ and $_\#\colon [X,\Omega Y] \longrightarrow [\Sigma X,Y]$).

0.4. Pullbacks, pushouts and Eckmann-Hilton duality

Unless explicitly stated otherwise, pullbacks and pushouts are homotopy (or weak) pullbacks and pushouts: If $f_i\colon X_i \to Y$, $i = 0,1$, *the pullback of* f_0, f_1 is the space W_{f_0,f_1} of all triples x,γ,y, $x \in X_0$, $\gamma\colon I \to Y$, $y \in X_1$ satisfying $\gamma(0) = f_0(x)$, $\gamma(1) = f_1(y)$ together with the two maps $g_i\colon W_{f_0,f_1} \to X_i$, $g_0(x,\gamma,y) = x$, $g_1(x,\gamma,y) = y$. The topology of W_{f_0,f_1} is the one inherited from $X_0 \times Y^I \times X_1$.

The pushout of $f_i\colon X \to Y_i$, $i = 0,1$, is the quotient space M_{f_0,f_1} of $Y_0 \vee X \times I \vee Y_1$ induced by the equivalence relation spanned by

$$f_0(x),*,* \equiv *,(x,0),*, \quad *,*,f_1(x) \equiv *,(x,1),*$$

$$*,(\ ,t),* \equiv *,\ *,\ * = *$$

A pushout has two structural maps $g_i\colon Y_i \to M_{f_0,f_1}$ induced by the inclusions $Y_i \subset Y_0 \vee X \times I \vee Y_1$.

If $f: X \to Y$ then $W_{*,f}$ is called the fiber of f, $M_{*,f} = C(f)$ is the cone on f. (If $1_X: X \to X$ is the identity map then $C(1_X) = CX$ - the cone on X. If $*: * \to X$ then $M_{*,*} = \Sigma X$.)

Pullbacks and pushouts have the following semi-universal properties: The diagrams

$$\begin{array}{ccc} W_{f_0,f_1} & \xrightarrow{g_1} & X_1 \\ \downarrow{\scriptstyle g_0} & & \downarrow{\scriptstyle f_1} \\ X_0 & \xrightarrow{f_0} & Y \end{array} \qquad\qquad \begin{array}{ccc} X & \xrightarrow{f_0} & Y_0 \\ \downarrow{\scriptstyle f_1} & & \downarrow{\scriptstyle g_0} \\ Y_1 & \xrightarrow{g_1} & M_{f_0,f_1} \end{array}$$

are commutative and for any space L and maps $\tilde{g}_i: L \to X_i$, $f_0\tilde{g}_0 \sim f_1\tilde{g}_1$ ($\tilde{g}_i': Y_i \to L$, $\tilde{g}_0'f_0 \sim \tilde{g}_1 f_1$) there exists a (non unique!) map $h: L \to W_{f_0,f_1}$ ($h': M_{f_0,f_1} \to L$) so that $\tilde{g}_i \sim g_i h$ ($\tilde{g}_i' \sim h' g_i$).

One can easily find duality properties between pullbacks and pushouts. This duality principal is referred to as *the Eckmann-Hilton duality* by which one interchanges pullbacks and pushouts, actions and coactions, principal fibration and principal cofibrations, Eilenberg-MacLane spaces and Moore spaces, homotopy groups and cohomology groups, etc. Some geometric proofs can be dualized to obtain the Eckmann-Hilton dual statements.

0.5. Ω-Spectra, ring spectra, generalized cohomology

An Ω-spectrum is a sequence of spaces and maps $E_* = \{E_n, \gamma_n\}$ where $\gamma_n: E_n \to \Omega E_{n+1}$ is a homotopy equivalence. γ_n induces a homotopy associative and homotopy commutative multiplication μ_n on E_n.

$\{E^n = [\ ,E_n]\}$ represents a reduced generalized cohomology theory. $([X,E_n] = E^n(X))$.

A ring structure for $\{E_n,\gamma_n\}$ are maps $\phi_{n,m}\colon E_n \wedge E_m \longrightarrow E_{n+m}$ so that the following four diagrams commute:

$L^{(S)}{}_1$

$$\begin{array}{ccccc} E_{n-1} \wedge E_m & \xrightarrow{\gamma_{n-1} \wedge 1} & \Omega E_n \wedge E_m & \xrightarrow{\hat{u}_1} & \Omega(E_n \wedge E_m) \\ \downarrow{\scriptstyle \phi_{n-1,m}} & & & & \downarrow{\scriptstyle \Omega\phi_{n,m}} \\ E_{n+m-1} & & \xrightarrow{\gamma_{n+m-1}} & & \Omega E_{n+m} \end{array}$$

$R^{(S)}{}_1$

$$\begin{array}{ccccc} E_n \wedge E_{m-1} & \xrightarrow{1 \wedge \gamma_{m-1}} & E_n \wedge \Omega E_m & \xrightarrow{\hat{u}_2} & \Omega(E_n \wedge E_m) \\ \downarrow{\scriptstyle \phi_{n,m-1}} & & & & \downarrow{\scriptstyle \Omega\phi_{n,m}} \\ E_{n+m-1} & & \xrightarrow{\gamma_{n+m-1}} & & \Omega E_{n+m} \end{array}$$

$\hat{u}_1\colon \Omega E_n \wedge E_m \longrightarrow \Omega(E_n \wedge E_m)$, $\hat{u}_2\colon E_n \wedge \Omega E_m \longrightarrow \Omega(E_n \wedge E_m)$ are the natural maps given by

$$(\hat{u}_1)_\# = (1)_\# \wedge 1, \qquad (\hat{u}_2)_\# = 1 \wedge (1)_\#$$

$$\hat{u}_1(\lambda,x)[t] = \lambda(t),x \qquad \hat{u}_2(x,\lambda)[t] = x,\lambda(t).$$

$L^{(S)_2}$

$$\begin{array}{ccc}
(E_n \times E_n) \wedge E_m & \xrightarrow{\ {}_R\widetilde{\Delta}\ } & (E_n \wedge E_m) \times (E_n \wedge E_m) \\
 & & \Big\downarrow {\scriptstyle \phi_{n,m} \times \phi_{n,m}} \\
\Big\downarrow {\scriptstyle \mu_n \wedge 1} & & E_{n+m} \times E_{n+m} \\
 & & \Big\downarrow {\scriptstyle \mu_{n+m}} \\
E_n \wedge E_m & \xrightarrow{\ \phi_{n,m}\ } & E_{n+m}
\end{array}$$

$R^{(S)_2}$

$$\begin{array}{ccc}
E_n \wedge (E_m \times E_m) & \xrightarrow{\ {}_L\widetilde{\Delta}\ } & (E_n \wedge E_m) \times (E_n \wedge E_m) \\
 & & \Big\downarrow {\scriptstyle \phi_{n,m} \times \phi_{n,m}} \\
\Big\downarrow {\scriptstyle 1 \wedge \mu_m} & & E_{n+m} \times E_{n+m} \\
 & & \Big\downarrow {\scriptstyle \mu_{n+m}} \\
E_n \wedge E_m & \xrightarrow{\ \phi_{n,m}\ } & E_{n+m}
\end{array}$$

${}_R\widetilde{\Delta}[(x,y),z] = [(x,z),(y,z)]$, ${}_L\widetilde{\Delta}[x,(y,z)] = [(x,y),(x,z)]$.

$\sigma\colon E^n(X) \to E^{n-1}(\Omega X)$ is given as follows: Let $x \in E^n(X)$ be represented by $f\colon X \to E_n$ then $\sigma x \in E^{n-1}[\Omega X]$ is represented by g, $\gamma_{n-1} g \sim \Omega f$.

If $\{E_n, \gamma_n, \phi_{n,m}\}$ is an Ω ring spectrum, then the function $\hat{\alpha}\colon [X,E_n] \times [Y,E_m] \longrightarrow [X \wedge Y, E_{n+m}]$ given by $\hat{\alpha}(f,g) = \phi_{n,m}(f \wedge g)$ is bilinear and induces a homomorphism

$$\alpha\colon [X,E_n] \otimes [Y,E_m] \longrightarrow [X \wedge Y, E_{n+m}].$$

The composition $[X,E_n] \otimes [X,E_m] \xrightarrow{\alpha} [X \wedge X, E_{n+m}] \xrightarrow{E^{n+m}(\bar{\Delta})} [X,E_{n+m}]$ induces a graded ring structure on $E^*(X)$ (not necessarily associative or commutative or with unit in its non-reduced version).

CHAPTER I

The Category of H-spaces

Introduction

This chapter is devoted to the study of the most elementary properties of H-spaces. With the exception of 1.1.3 and 1.2.3 all proofs are given in details. Only the most fundamental Homotopy Theory is used.

It is very difficult to trace the origin of many statements. Some references are given but there is no certainty that these are the earliest. Other statements should be considered as "folklore" and other appear here possibly for the first time.

Section 1 contains observations which follow directly from the definitions of H-spaces. It contains a review of the notion of the Moore-Path space which replaces throughout these notes the ordinary space of paths and is used in describing homotopies.

Section 2 is devoted to a preliminary study of special classes of H-spaces such as homotopy commutative and homotopy associative H-spaces with some examples.

The algebraic properties of the set of homotopy classes of maps into an H-space is studied in Section 3. Some of the notations used in the sequel are established here. The notions studied in Section 3 are used in Section 4 to define the first obstructions in the theory of H-spaces. In this section the problem of enumerating the H-structures on a given space is briefly discussed.

Section 5 is devoted to some analysis of the obstructions for a map to be an H-map and ways for its anihilation.

1.1. Basic properties of H-spaces

An H-space is a pair X,μ where X is a space and $\mu: X \times X \to X$ satisfies the properties $\mu(x,*) = x = \mu(*,x)$.

Let $F: X \vee X \to X$ be defined by $F(x,*) = x = F(*,x)$ (F is called "the folding map"). Thus, an H-space X,μ is a space X with a map $\mu: X \times X \to X$ so that $\mu|X \vee X = F$. If X,μ is an H-space we call μ a multiplication or an H-structure for X. Thus, an H-space is a space together with a continuous multiplication with a unit. From the homotopy theory point of view one may replace the unit by a homotopy unit, i.e.: in the definition of an H-space replace the property $\mu|X \vee X = F$ by the requirement $\mu|X \vee X \sim F$. However, with our notion of a space by the homotopy extension property a multiplication with a homotopy unit can be homotoped to a multiplication with strict unit. The latter will be the only type of multiplication considered in these notes.

Two examples of H-spaces come in mind: Topological groups and the spaces of loops. The first has a strict unit the other has a homotopy unit. Later in this section we shall introduce its equivalent - one with a strict unit, namely the Moore-Loop Space.

Our first simple observation deals with homotopy-groups type functors applied to H-spaces:

1.1.1. Proposition [Hilton]: *Let* π *be a functor from the category of spaces and homotopy classes of maps into the category of abelian groups which preserves products, i.e.;*

$$\alpha: \quad \pi(X \times Y) \approx \pi(X) \oplus \pi(Y), \quad \pi(f \times g) = \pi(f) \oplus \pi(g)$$

and $\pi(*) = 0$.

If X,μ *is an H-space then*

$$\pi(\mu)\alpha^{-1}\colon\quad \pi(X) \oplus \pi(X) \to \pi(X)$$

coincides with the group addition $x,y \to x+y$.

<u>Proof</u>: Let $i_1\colon X \to X \times X$, $i_2\colon X \to X \times X$ be given by $i_1(x) = x,*$ $i_2(x) = *,x$. Then $j_1 = \alpha\pi(i_1)\colon \pi(X) \to \pi(X) \oplus \pi(X)$ is the homomorphism $x \to x,0$, $j_2(x) = 0,x$ and $x,y = j_1(x) + j_2(y)$.

$$\pi(\mu)\alpha^{-1}(x,y) = \pi(\mu)\alpha^{-1}(j_1x + j_2y) = \pi(\mu)\pi(i_1)x + \pi(\mu)\pi(i_2)y$$

But $\mu i_1 = 1$ $\mu i_2 = 1$, hence $\pi(\mu)\pi(i_k) = 1$, $k = 1,2$, and $\pi(\mu)\alpha^{-1}(x,y) = x+y$.

<u>1.1.2. Lemma (see [Copeland]$_1$ and [Croon]:</u>

(a) X *admits an H-structure if and only if for every space* Y $[Y,X]$ *admits a multiplication with unit in a natural way, i.e.* $[\ ,X]$ *is a functor into the category of sets with multiplication with units.*

(b) X *admits an H-structure if and only if for every pair of spaces* M,L

$$i^*\colon\quad [M \times L,X] \to [M \vee L,X]$$

is surjective where i^* *is induced by the inclusion* $i\colon M \vee L \to M \times L$.

(c) If X *is an H-space there exists a homotopy equivalence* α*:*

$$\begin{array}{ccc}
\hat{\Omega}X & \xrightarrow{\quad 1 \quad} & \hat{\Omega}X \\
\downarrow{\scriptstyle i_2} & & \downarrow{\scriptstyle i} \\
X \times \hat{\Omega}X & \underset{\approx}{\xrightarrow{\quad \alpha \quad}} & X^{S^1} \\
\downarrow{\scriptstyle p_1} & & \downarrow{\scriptstyle E} \\
X & \xrightarrow{\quad 1 \quad} & X
\end{array}$$

where $X^{S^1} = \{f\colon I \to X \mid f(0) = f(1)\}$,

$\hat{\Omega}X = \{f\colon I \to X \mid f(0) = f(1) = *\}$, $E(f) = f(0)$, $i_2(\lambda) = *,\lambda$

and i *- the inclusion.* $\alpha|X = \chi\colon X \to X^{S^1}$ *given by*

$\chi(x)[t] = x$ for every t. *(And see* 1.3.6 *in the sequel.)*

(d) A retract of an H-space is an H-space.

Proof: (a) If $\mu\colon X \times X \to X$ is a multiplication then $[Y,X] \times [Y,X] \xrightarrow{\approx} [Y,X \times X] \xrightarrow{\mu_*} [Y,X]$ is a multiplication with $*\colon Y \to * \to X$ as a unit.

If $[Y,X]$ has a multiplication $\cdot$ with a unit for every Y and if $p_i\colon X{\times}X \to X$ $i = 1,2$ are the projections then a map representing $[p_1]\cdot[p_2] \in [X \times X,X]$ is a multiplication for X: Indeed, as $[*,X]$ is a singleton $[*,X] = 1$ and $* \in [Y,X]$ must be the unit. If $i_k\colon X \to X \times X$ $k = 1,2$ are the injections $p_1 i_1 = 1$ $p_2 i_1 = *$ and $([p_1] \cdot [p_2])[i_1] = [1] \cdot * = [1]$, similarly for i_2.

(b) Suppose $\mu\colon X \times X \to X$ is a multiplication. Let $f\colon M \vee L \to X$. Put $f_M = f|M$, $f_L = f|L$. Put $\hat{f} = \mu(f_M \times f_L)$. Then $\hat{f}$ extends f: $i^*[\hat{f}] = [f]$, and i^* is a surjection.

If $i^*\colon [M \times L,X] \to [M \vee L,X]$ is a surjection for every M,L then for $M = L = X$ any element in $(i^*)^{-1}[F]$ is a multiplication.

(c) Let $\alpha\colon X \times \hat{\Omega}X \to X^{S^1}$ be given by $\alpha(x,\varphi)[t] = \mu(x, \varphi(t))$ where μ is a multiplication. $\alpha(x,*)[t] = x$ implies $\alpha|X = \chi$. $E\alpha(x,\varphi) = \alpha(x,\varphi)[0] = \mu(x, \varphi(0)) = \mu(x,*) = x = p_1(x,\varphi)$ and hence $E\alpha = p_1$.

If $\lambda \in \hat{\Omega}X$ then $\alpha i_2(\lambda)[t] = \mu(*, \lambda(t)) = \lambda(t)$ and $\alpha i_2 = i$. As p_i and E are fibrations the exact sequence of homotopy groups and the five lemma imply that $\pi(\alpha)$ is an isomorphism and consequently α is a homotopy equivalence.

(d) Let $i\colon A \subset X$ be a retract with a retraction $\tau\colon X \to A$. If $\mu\colon X \times X \to X$ is a multiplication then so is $\tau\mu(i\times i)\colon A \times A \to A$.

<u>1.1.3. Theorem [James]$_1$</u>: *For a space* X *let* $\sigma\colon X \to \hat{\Omega}\Sigma X$ *be the adjoint of* $1\colon \Sigma X \to \Sigma X$ *where* $\Sigma X = S^1 \wedge X = S^1 \times X/S^1 \vee X$. *I.e.* $\sigma(x)[t] = [x,t]$.

X *is an H-space if and only if* σ *has a homotopy left inverse.*

<u>Proof</u>: See [James]$_1$.

<u>1.1.4. Corollary</u>: *Let* $f\colon A \to B$. $f^*\colon [B,X] \to [A,X]$ *is surjective for every H-space* X *if and only if* $\Sigma f\colon \Sigma A \to \Sigma B$ *has a homotopy left inverse.*

<u>Proof</u>: Suppose $f^*\colon [B,X] \to [A,X]$ is surjective for all H-spaces X. Then $\sigma\colon A \to \hat{\Omega}\Sigma A$ can be f-extended, i.e.: there exists $h\colon B \to \hat{\Omega}\Sigma A$, $hf \sim \sigma$. If $(\ )_\#$ is the adjoint operation then
$(h)_\#(\Sigma f) = (hf)_\# \sim \sigma_\# = 1$ and $(h)_\#$ is a homotopy left inverse of Σf.

Suppose Σf has a left inverse $\hat{h}$. Given $g\colon A \to X$. Consider $A \xrightarrow{g} X \xrightarrow{\sigma} \hat{\Omega}\Sigma X$ and the adjoint $(\sigma g)_\#\colon \Sigma A \to \Sigma X$. $\hat{g} = (\sigma g)_\#\hat{h}\colon \Sigma B \to \Sigma X$ is a Σf extension of $(\sigma g)_\#$: $\hat{g}\Sigma f = (\sigma g)_\#\hat{h}\Sigma f \sim (\sigma g)_\#$. Taking an adjoint of $\Sigma A \xrightarrow{\Sigma f} \Sigma B \xrightarrow{\hat{g}} \Sigma X$ one gets $A \xrightarrow{f} B \xrightarrow{\hat{g}\#} \hat{\Omega}\Sigma X$, $\hat{g}_\# f = \sigma g$ so $\hat{g}_\#$ f-extends σg. If X is an H-space by 1.1.3 there exists $r\colon \Omega\Sigma X \to X$ $r\sigma \sim 1$ and $r\hat{g}_\#\colon B \to X$ f extends g: $r\hat{g}_\# f \sim r\sigma g \sim g$.

<u>1.1.4.1. Remark</u>: We use here $\hat{\Omega}$ to denote the loop space instead of the customary Ω in order to preserve the latter to denote the Moore loop space which is going to replace loops throughout these notes.

<u>1.1.5. Proposition [James]$_3$</u>: *Let* $A \subset B$ *and suppose* ΣA *is a retract of* ΣB. *Let* X,μ *be an H-space. Given maps* $g_0,g_1\colon B \to X$. *If*

$g_0|A = g_1|A$ *and* $g_0 \sim g_1$ *then* $g_0 \sim g_1$ rel A, *i.e.: There exists a homotopy* $F\colon B \times I \to X$, $F(b,0) = g_0(b)$, $F(b,1) = g_1(b)$ *and* $F(a,t) = g_0(a) = g_1(a)$ *for every* $a \in A$.

To prove 1.1.5 we first prove:

<u>1.1.5.1. Lemma</u>: *Let* $A \subset B$ *and let* Y *be any space. Then there exists a homotopy equivalence* $u\colon \Sigma(B \cup (A \times Y)) \xrightarrow{\approx} \Sigma B \vee \Sigma A \wedge Y \vee \Sigma Y$ *which is natural with respect to maps* $B,A \to B',A'$ *and* $Y \to Y'$. *In particular, if* $\Sigma A \subset \Sigma B$ *is a retract then* $\Sigma(B \cup A \times Y) \subset \Sigma(B \times Y)$ *is a retract.*

<u>Proof</u>: Define maps $v_1\colon B \cup (A \times Y) \to B$, $v_2\colon B \cup (A \times Y) \to A \wedge Y$ and $v_3\colon B \cup (A \times Y) \to Y$ by:

$$v_1(b) = b \quad v_1(a,y) = a$$

$$v_2(b) = * \quad v_2(a,y) = \Lambda(a,y)$$

$$v_3(b) = *, \quad v_3(a,y) = y$$

v_i are obviously natural with respect to maps $B,A \to B',A'$ and $Y \to Y'$. Define $u\colon \Sigma(B \cup A \times Y) \to \Sigma B \vee \Sigma A \wedge Y \vee \Sigma Y$ by

$$u(t,x) = \begin{cases} [3t, v_1(x)], \quad *, \quad * & 0 \le t \le \frac{1}{3} \\ *, \quad [3t-1, v_2(x)], \quad * & \frac{1}{3} \le t \le \frac{2}{3} \\ *, \quad *, \quad [3t-2, v_3(x)] & \frac{2}{3} \le t \le 1 \end{cases}$$

Using the basic homology computations one can see that $H(v_1) \oplus H(v_2) \oplus H(v_3)\colon H(B \cup A \times Y) \to H(B) \oplus H(A \wedge Y) \oplus H(Y)$ is an isomorphism hence u is a homotopy equivalence.

One has the following (induced by $B,A \subset B,B$)

$$\begin{array}{ccc} \Sigma(B \cup A \times Y) & \xrightarrow[\approx]{u} & \Sigma B \vee \Sigma(A \wedge Y) \vee \Sigma Y \\ \Big\downarrow \hat{\imath} & & \Big\downarrow 1 \vee \Sigma i \wedge 1 \vee 1 \\ \Sigma(B \times Y) & \xrightarrow[\approx]{u} & \Sigma B \vee \Sigma(B \wedge Y) \vee \Sigma Y \end{array}$$

where $i: A \subset B$. If Σi has a left inverse so does $\hat{\imath}$ and $\Sigma[B \cup A \times Y]$ is a retract of $\Sigma(B \times Y)$.

<u>Proof of 1.1.5</u>: Let $\hat{F}: B \times I \to X$ be any homotopy of g_0 and g_1. As $g_0|A = g_1|A$, $\hat{F}|A \times I$ factors through $\hat{F}_0: A \times S^1 \to X$. ($S^1 = \{z \mid |z| = 1\}$, $* = 1$). $\hat{F}_0$ easily extends to $\hat{F}_1: B \cup A \times S^1 \to X$ by putting $\hat{F}_1(b) = g_1(b)$ (and note that for $a \in A$ $g_1(a) = \hat{F}_1(a,*) = g_0(a) = g_1(a)$). By 1.1.4 and 1.1.5.1 $\hat{F}_1$ extends to $\hat{F}_2: B \times S^1 \to X$. Define $\hat{F}_3: B \times I \to X$ by

$$\hat{F}_3(b,t) = \begin{cases} \hat{F}(b, 2t) & 0 \le t \le \frac{1}{2} \\ \hat{F}_2(b, e^{4\pi i(1-t)}) & \frac{1}{2} \le t \le 1 \end{cases}$$

$$\hat{F}(b,1) = g_1(b) = \hat{F}_2(b,1) = \hat{F}_2(b,*)$$

$$\hat{F}_3(b,0) = \hat{F}(b,0) = g_0(b). \quad \hat{F}_3(b,1) = \hat{F}_2(b,*) = g_1(b).$$

Hence $\hat{F}_3$ is a homotopy $g_0 \sim g_1$. For $a \in A$

$$\hat{F}_3(a,t) = \begin{cases} \hat{F}(a, 2t) & 0 \le t \le \frac{1}{2} \\ \hat{F}(a, 2(1-t)) & \frac{1}{2} \le t \le 1 \end{cases}$$

hence $\hat{F}_3|A \times I \sim p_1$ rel $A \times \dot{I}$ and by standard homotopy theory $\hat{F}_3 \sim F$ rel $B \times \dot{I}$ and $F|A \times I = p_1$. $F: g_0 \sim g_1$ rel A.

Using arguments similar to those of 1.1.5.1 one has

<u>1.1.6. Examples</u>: Given spaces $A_1, A_2, \ldots, A_k$. Let $\bigvee_{i=1,\ldots,k}^{n} A_i \subset \prod_{i=1}^{k} A_i$

$\bigvee_{i=1,\ldots,k}^{n} A_i = \{(a_1, a_2, \ldots, a_k) \in \prod_{i=1}^{k} A_i \mid a_j = *$ for at least $k - n$ indices $j\}$. Then $\Sigma(\bigvee_{i=1,\ldots,k}^{n} A_i) \to \Sigma(\prod_{i=1}^{k} A_i)$ has a left inverse.

If X,μ and X',μ' are H-spaces a $\mu - \mu'$ H-map $f: X,\mu \to X',\mu'$ is a map $f: X \to X'$ so that $f\mu \sim \mu' f \times f$. In view of 1.1.5 and 1.1.6, as $f\mu | X \vee X = fF = Ff \vee f = \mu'(f \times f) | X \vee X$ one may always assume that $f\mu \sim \mu' f \times f$ rel $X \vee X$.

The class of H-spaces and H-maps forms a category H_f.

Sometimes it is convenient to carry the homotopy $F: f\mu \sim \mu' f \times f$ as a structural part of an H-map. In order to do this in a managable way one should recall the notion of the Moore path space (and see e.g: [Fuchs]):

Let R^+ be the set of non negative real numbers. Given a space X the Moore path space of X is the space

$$PX = \{t, \varphi \in R^+ \times X^{R^+} \mid \varphi(s) = \varphi(t) \quad \text{if} \quad s \geq t\}.$$

We give PX the union (or weak) topology induced by the inclusions

$$PX \cap ([0,a] \times X^{[0,a]}) \subset PX.$$

We call $\ell: PX \to R^+$, $\ell(t,\varphi) = t$, the length function. Define the evaluation at s map $(0 \leq s \leq \infty)$ $E_s: PX \to X$ by $E_s(t,\varphi) = \varphi(s)$ if $s < \infty$ and $E_\infty(t,\varphi) = \varphi(t)$. Let $k_t: X \to PX$ be the length t constant path map

$$k_t(x) = (t, \hat{k}(x)) \quad \hat{k}(x)(s) = x \quad \text{for all} \quad s.$$

We write $k_0 = k$.

P is a functor from the category of spaces and maps into itself: If $f\colon X \to Y$ then $Pf\colon PX \to PY$ is given by $Pf(t,\varphi) = t,\ f\varphi$. One has homotopy equivalences $a\colon P(X \times Y) \to PX \times PY$ and $b\colon PX \times PY \to P(X \times Y)$. $a(t,\varphi) = (t,\ p_1\varphi),\ (t,\ p_2\varphi)$ where $p_1\colon X \times Y \to X$, $p_2\colon X \times Y \to Y$ are the projections. $b[(t_1,\varphi_1),\ (t_2,\varphi_2)] = \max(t_1,t_2),\ (\varphi_1 \times \varphi_2)\Delta_{R^+}$, $ba = 1$ and $(ab) \sim 1$. If $h\colon X_1 \times X_2 \to Y$ we shall write $Ph\colon PX_1 \times PX_2 \to PY$ for Phb. Let $LX = E_0^{-1}(*)$, $\Omega X = E_0^{-1}(*) \cap E_\infty^{-1}(*)$. $P_2X \subset PX \times PX$ is the set of pairs $P_2X = \{(\overline{\varphi}_1,\ \overline{\varphi}_2) \in PX \times PX \mid E_\infty(\overline{\varphi}_1) = E_0(\overline{\varphi}_2)\}$.

The path addition map $\text{Add}\colon P_2X \to PX$ is defined by

$$\text{Add}[(t_1,\ \varphi_1),\ \ (t_2,\ \varphi_2)] = t_1 + t_2,\ \varphi_{12}$$

and φ_{12} is defined

$$\varphi_{12}(s) = \begin{cases} \varphi_1(s) & 0 \leq s \leq t_1 \\ \varphi_2(s-t_1) & t_1 \leq s \end{cases}$$

Write $\text{Add}(\overline{\varphi}_1,\ \overline{\varphi}_2) = \overline{\varphi}_1 + \overline{\varphi}_2$ then whenever the following are defined they satisfy

$$\overline{\varphi}_1 + (\overline{\varphi}_2 + \overline{\varphi}_3) = (\overline{\varphi}_1 + \overline{\varphi}_2) + \overline{\varphi}_3$$

and always $k(E_0(\overline{\varphi})) + \overline{\varphi} = \overline{\varphi} = \overline{\varphi} + k(E_\infty(\overline{\varphi}))$.

Let $c\colon PX \to PX$ be given by $c(t,\varphi) = t,\hat{\varphi}$ where $\hat{\varphi}$ is defined by

$$\hat{\varphi}(s) = \begin{cases} \varphi(t-s) & 0 \leq s \leq t \\ \varphi(0) & t \leq s \end{cases}$$

We write $-\overline{\varphi}$ for $c(\overline{\varphi})$.

Note that $\Omega X \times \Omega X \subset \Omega X \times LX \subset P_2X$ and that

$$\mathrm{Add}(\Omega X \times \Omega X) \subset \Omega X$$

$$\mathrm{Add}(\Omega X \times LX) \subset LX$$

Hence Add defines an associative multiplication with unit on ΩX and an action of ΩX on LX. One has a homotopy equivalence $X^I \to PX$ given by $\varphi \to 1,\varphi'$ $\varphi'(s) = \varphi(s)$ for $s \leq 1$ and $\varphi'(s) = \varphi(1)$ for $s \geq 1$. The inverse homotopy equivalence can be given by $PX \to X^I$, $t,\varphi \to \varphi_t$ where

$$\varphi_t(s) = \begin{cases} \varphi(t\cdot s) & 1 \leq t \\ \varphi(s) & t \leq 1 \end{cases}$$

$PX \to X^I \to PX$ and $X^I \to PX \to X^I$ are homotopic to the identity via homotopies corel the evaluations at the endpoints.

A homotopy $M \times I \to X$ or its adjoint $M \to X^I$ can be replaced by a map $M \to PX$. Thus, a homotopy $F\colon f_1 \sim f_2$, $f_i\colon M \to X$ is a map $F\colon M \to PX$ with $E_0F = f_1$, $E_\infty F = f_2$.

If $g_1, g_2\colon X \to Y$, $G\colon X \to PY$ a homotopy $G\colon g_1 \sim g_2$, then there exists $\hat{G}\colon PX \to P(PY)$ $\hat{G}\colon Pg_1 \sim Pg_2$. G satisfies

$$PE_s\hat{G} = GE_s \quad \text{and} \quad E_sE_t\hat{G} = E_tGE_s.$$

Let $F_1\colon M \times I \to X$ be a homotopy $f_1 \sim f_2$ rel $L \subset M$. Let $\eta\colon M \to R^+$ satisfy $\eta^{-1}(0) = L$. Consider the function $\hat{F}\colon M \times R^+ \to X$ given by

$$\hat{F}(x,t) = \begin{cases} F_1(x, \frac{t}{\eta(x)}) & x \notin L, \quad t \leq \eta(x) \\ F_1(x,1) & x \notin L, \quad t \geq \eta(x) \\ f_1(x) = f_2(x) & x \in L \end{cases}$$

Then $\hat{F}$ is continuous and corresponds to a map $F: M \to PX$, $\ell F = \eta$ and $F|L = kf_1|L = kf_2|L$.

Thus a homotopy $f: M \to PX$, $F: f \sim g$ is rel L if $F|L = kf$. As all homotopies are rel $*$ we always assume $F(*) = k(*)$.

Given $h: X \to Y$, $f_0, f_1: M \to X$. A homotopy $F: M \to PX$, $F: f_0 \sim f_1$, is corel h if $PhF = k_{\ell F}hf_0 = k_{\ell F}hf_1$.

Now 1.1.5 can be generalized:

1.1.7. Proposition: *Suppose* $A \subset B$, ΣA *a retract of* ΣB. *Let* X,μ *be an H-space. Given* $g_1, g_2: B \to X$, $g_1 \sim g_2$. *Then any homotopy* $g_1|A \sim g_2|A$ *can be extended to a homotopy* $g_1 \sim g_2$.

Proof: Given $F_0: A \to PX$, $\tilde{F}_0: g_1|A \sim g_2|A$. By the homotopy extension property $\tilde{F}_0$ extends to $\hat{F}_0: B \to PX$, $E_0\hat{F}_0 = g_1$, $E_\infty\hat{F}_0 = g_2'$. One has $g_2' \sim g_2$ and $g_2|A = g_2'|A$. By 1.1.5 there exists $\hat{F}_1: g_2' \sim g_2$ rel A. $\hat{F}_0 + \hat{F}_1: g_1 \sim g_2$ extends F_0.

If $X,\mu \to X',\mu'$ is an H-map one has $f\mu \sim \mu'(f \times f)$ rel $X \vee X$. Hence, there exists $F: X \times X \to PX'$, $F|X \vee X = kF$, $E_0F = f\mu$, $E_\infty F = \mu(f \times f)$.

1.1.8. Definition: *The precise category of H-spaces* is the category H_{fF} whose objects are H-spaces X,μ and the morphisms $X,\mu \to X',\mu'$ are pairs f, F, $f: X \to X'$, $F: X \times X \to PX'$, $F|X \vee X = kfF$, $E_0F = f\mu$, $E_\infty F = \mu'(f \times f)$.

The composition is given by

$$(f',F')(f,F) = f'f, \quad Pf'F + F'(f \times f).$$

$1_{X,\mu} = 1_X, \; k\mu$.

1.1.8.1. Remark: Throughout these notes we shall use Moore path spaces (based paths and loops) rather than the ordinary paths (based paths and loops) and all homotopies will be considered as maps into Moore path spaces.

1.1.9. Remark: The use of the precise category will be limited in the sequel only to the cases where the homotopy F is specifically needed and we shall use the notations $f\colon X,\mu \to X',\mu'$ and $f,F\colon X,\mu \to X',\mu'$ for H-maps as the case requires.

1.2. Some special classes of H-spaces

It was observed in the case of a loop space that a multiplication may possess some additional properties, such as associativity and a weak form of an inverse. From the homotopy point of view these are defined as follows:

1.2.1. Definition: An H-space X,μ is said to be *homotopy associative* (an HA space) if $\mu(\mu \times 1) \sim \mu(1 \times \mu)$, i.e: If we denote $\mu(x,y) = x \cdot y$ then X,μ is an HA space if the maps $x,y,z \to (x\cdot y)z$ and $x,y,z \to x(y\cdot z)$ are homotopic.

X,μ is said to be *homotopy commutative* (an HC space) if $\mu \sim \mu T$ where $T\colon X \times X \to X \times X$ is the twisting $T(x,y) = y,x$, i.e: X,μ is HC if the maps $x,y \to x\cdot y$ and $x,y \to y\cdot x$ are homotopic.

A *right(left) homotopy inverse* for X,μ is a map $C_r\colon X \to X$ ($C_\ell\colon X \to X$) satisfying

$$\mu(1 \times C_r)\Delta \sim * \qquad (\mu(C_\ell \times 1)\Delta \sim *)$$

i.e: $C_r(C_\ell)$ satisfy: The maps $x \to x\cdot C_r x$ ($x \to C_\ell x\cdot x$) are null homotopic.

1.2.2. Examples: As was already observed in section 1.1 ΩX, Add is an HA space. All topological groups are HA.

S^7 regarded as the unimodular Cayley numbers is an H-space but not HA ([James]$_2$).

Among finite CW H-spaces we refer to compact Lie groups and S^7 as the classical H-spaces. All but S^7 are HA. In the sequel we shall introduce some non classical finite CW H-spaces, some of them will be HA and some will not. As far as HC one has the following interesting fact

1.2.3. Theorem [Hubbuck]$_2$: *All finite CW H-spaces which are HC spaces are of the homotopy type of a torus* $T^n = \underbrace{S^1 \times S^1 \times \ldots \times S^1}_{n}$ $n \geq 0$ $(T^0 = *)$. *Hence, there are no non classical finite CW HC spaces.*

If X,μ is an H-space $P\mu\colon PX \times PX \to PX$, the pointwise multiplication of paths, is defined by

$$P\mu[(t_1,\varphi_1),\ (t_2,\varphi_2)] = \max(t_1,t_2),\ \varphi_1\cdot\varphi_2$$

$$\varphi_1\cdot\varphi_2(s) = \mu(\varphi_1(s),\ \varphi_2(s)).$$

Thus, for pairs in P_2X (X - an H-space) one has two different binary operations - $P\mu$ and Add. The relation between them is given by

1.2.4. Lemma: *Let* $w_i\colon PX \times PX \to PX \times PX$ $i = 1,2$ *be given by*

$$p_1w_1 = P\mu(1 \times k_0E_0) \qquad p_2w_1 = P\mu(k_0E_\infty \times 1)$$

$$p_1w_2 = P\mu(k_0E_0 \times 1) \qquad p_2w_2 = P\mu(1 \times k_0E_\infty)$$

Then $\operatorname{im} w_i \subset P_2X$ *and* $P\mu \sim \operatorname{Add} w_i$.

Proof:

$$E_\infty p_1 w_1 = \mu(E_\infty \times E_0) = E_0 p_2 w_1$$

$$E_\infty p_1 w_2 = \mu(E_0 \times E_\infty) = E_0 p_2 w_2$$

Hence, $\mathrm{im}\, w_i \subset P_2X$. Let $\alpha_1 : PX \times PX \to PX \times PX$ be given by $\alpha_1(\overline{\varphi}_1,\overline{\varphi}_2) = \overline{\varphi}_1 + k_{\ell(\overline{\varphi}_2)}(E_\infty(\overline{\varphi}_1)),\ k_{\ell(\overline{\varphi}_1)}(E_0(\overline{\varphi}_2)) + \overline{\varphi}_2$. Then $\alpha_1 \sim 1$ corel $E_i \times 1$, $1 \times E_j$ $i = 0,\infty$, $j = 0,\infty$, and $P\mu\alpha_1 = \mathrm{Add}w_1$.
Similarly if

$$\alpha_2(\overline{\varphi}_1,\overline{\varphi}_2) = k_{\ell(\overline{\varphi}_2)}(E_0(\overline{\varphi}_1)) + \overline{\varphi}_1,\quad \overline{\varphi}_2 + k_{\ell(\overline{\varphi}_1)}E_\infty(\overline{\varphi}_2)$$

$\alpha_2 \sim 1$ corel $E_i \times 1$, $1 \times E_i$ $i = 0,\infty$ and $P\mu\alpha_2 = \mathrm{Add}w_2$.

1.2.5. Corollary: *On* $\Omega X \times LX$ $P\mu \sim \mathrm{Add}$ corel E_i *and on* $LX \times \Omega X$ $P\mu \sim \mathrm{AddT}$ corel E_i, $i = 0,\infty$.

On $\Omega X \times \Omega X$ $\mathrm{Add} \sim \mathrm{AddT}$, *hence* $(\Omega X, \mathrm{Add})$ *is HC*.

Proof: On $\Omega X \times LX$ $w_1 = 1$, on $LX \times \Omega X$ $w_2 = T$.
On $\Omega X \times \Omega X = (\Omega X \times LX) \cap (LX \times \Omega X)$ $\mathrm{Add} \sim P\mu \sim \mathrm{AddT}$.

1.3. The structure of [, H-space].

As observed in 1.1.2(a), if X,μ is an H-space for any space M $[M,X]$ has a multiplication with unit

$$\mu_*: [M,X] \times [M,X] = [M,X \times X] \to [M,X].$$

It can be easily seen that if X,μ is HA (HC) μ_* is associative (commutative).

1.3.1. Theorem ([James]$_3$: The proof following [Stasheff]$_5$): *If* X,μ *is an H-space then for any space* M $[M,X]$ *is an algebraic loop, i.e: For any pair* $f,g \in [M,X]$ *there exists a unique* $D_{f,g} \in [M,X]$ *so that* $D_{f,g} + g = \mu_*(D_{f,g}, g) = f$.

Proof: Define $s: X \times X \to X \times X$ by $s(x,y) = \mu(x,y), y$. Then $\pi_n(s): \pi_n(X) \oplus \pi_n(X) \to \pi_n(X) \oplus \pi_n(X)$ is given by $\pi_n(s)(a,b) = a + b,b$.

Hence $\pi_n(s)$ is an isomorphism and s is a homotopy equivalence. Let $D_{f,g}: M \to X$ be the (unique) element so that $(D_{f,g}, g) = (D_{f,g} \times g)\Delta$ is the unique solution of the lifting problem

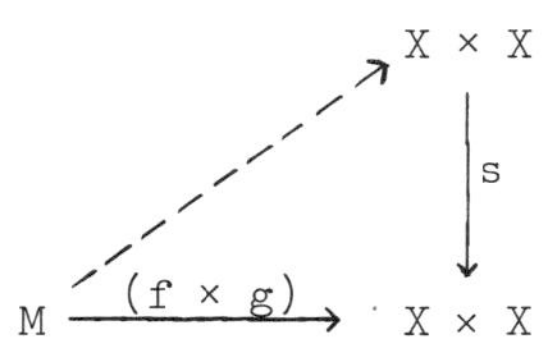

$D_{f,g}$ is the unique solution of the equation $\mu_*(x,g) = f$.

1.3.2. Corollary: *If* X,μ *is HA then* $[M,X]$ *is a group and* $D_{f,g} = f-g$. *If in addition* X *is HC then* $[M,X]$ *is an abelian group.*

1.3.3. Remark: Even though $[M,X]$ in general is not abelian, in order to distinguish between the operation in $[M,X]$ and the map-composition operation we use $+$ for the operation μ_*, $\mu_*(f,g) = f+g$.

The function $D: [K,X] \times [K,X] \to [K,X]$ is referred to as the difference and is uniquely defined by the property $f \sim \mu(D_{f,g} \times g)\Delta$.

The proof of 1.3.1 holds for a general space X and a CW complex M. However in our case one has a "universal" difference, namely

$$D = D_{p_1,p_2}: X \times X \to X$$

and $D_{f,g} = D(f \times g)\Delta_M$ for $f,g\colon M \to X$. The uniqueness of D implies the following:

$D\Delta = D_{1,1}\colon X \to X$ is the unique map satisfying $\mu(D_{1,1} \times 1)\Delta \sim 1$. As $\mu(* \times 1)\Delta = 1$, $D_{1,1} \sim *$. Similarly, if $i_1\colon X \to X \times X$ is given by $i_1(x) = x,*$ then $Di_1 = D_{1,*}$ satisfies $1 \sim \mu(D_{1,*} \times *)\Delta$. As $\mu(1 \times *)\Delta = 1$ $D_{1,*} \sim 1$. Now, $u\colon X \vee X \to X \times X$, $u(x,*) = x,*$ $u(*,x) = x,x$ is a cofibration. Note that if $s\colon X \times X \to X \times X$, $s(x,y) = \mu(x,y), y$, then $s|X \vee X = u$. Consequently one may assume that $D(x,*) = x$, $D(x,x) = *$, hence, $\mu(D \times 1)(1 \times \Delta)u = F$ and as Σu has a right inverse by 1.1.5 we have $\mu(D \times 1)(1 \times \Delta) \sim p_1$ rel u.

The uniqueness of D easily implies:

1.3.4. Lemma: *Let* $h\colon M \to L$, $g\colon X,\mu \to X',\mu'$ *be a map and an H-map respectively. For* $f_1,f_2 \in [L,X]$

$$gD_{f_1,f_2} = D_{gf_1,gf_2}, \qquad D_{f_1,f_2}h = D_{f_1h,f_2h}$$

1.3.5. Lemma (see [Copeland]$_1$ and [Arkowitz]): *If* X,μ *is an H-space then for any* K,L

$$0 \to [K \wedge L,X] \xrightarrow{\Lambda^*} [K \times L,X] \xrightarrow{i^*} [K \vee L,X] \to 0$$

is short exact in the following sense: i^* *is onto,* Λ^* *is* 1-1 *and* $i^*f = i^*g$ *if and only if* $D_{f,g} \in \operatorname{im} \Lambda^*$.

Proof: i^* is surjective by 1.1.2(b). Consider the Puppe sequence for an arbitrary space X

$$[\Sigma(K \times L),X] \xrightarrow{\Sigma i^*} [\Sigma(K \vee L),X] \xrightarrow{\partial} [K \wedge L,X] \xrightarrow{\Lambda^*} [K \times L,X]$$

as (by 1.1.6) Σi^* has a right inverse it is surjective and $\partial = *$.

Hence $\Lambda^{*-1}(*) = *$. If X,μ is an H-space $\Lambda^* h_1 = \Lambda^* h_2$ implies $\Lambda^* D_{h_1,h_2} = D_{h_1,h_2}\Lambda = D_{h_1\Lambda,h_2\Lambda} = *$. Hence $D_{h_1,h_2} = *$, $h_1 = h_2$ and Λ^* is 1-1.

Finally $i^* f_1 = i^* f_2$ if and only if $* = D_{f_1 i, f_2 i} = D_{f_1,f_2} i = i^* D_{f_1,f_2}$ and by the Puppe sequence this is equivalent to $D_{f_1,f_2} \in \operatorname{im} \Lambda^*$.

The following is a more precise version of 1.1.2(c). Here $X^{S^1} = \{\bar{\varphi} \in PX \mid E_0(\bar{\varphi}) = E_\infty(\bar{\varphi})\}$, $i\colon \Omega X \to X^{S^1}$ the inclusion, $E = E_0|X^{S^1} = E_\infty|X^{S^1}$.

1.3.6. Proposition: *Let* X,μ *be an H-space. Let* $\alpha\colon X \times \Omega X \to X^{S^1}$ *be given by* $\alpha = P\mu(i \times k)T$. *Let* $\Upsilon\colon X^{S^1} \to X \times \Omega X$ *be given by* $\Upsilon = [E \times PD(1 \times kE)\Delta]\Delta$. *Then* $\alpha\Upsilon \sim 1$ corel E, $\Upsilon\alpha \sim 1$ corel p_1.

Proof:

$$\alpha\Upsilon = [P\mu(i \times k)T][E \times PD(1 \times kE)\Delta]\Delta =$$

$$= P\mu[PD(1 \times kE) \times kE](1 \times \Delta)\Delta =$$

$$= P\mu[PD(1 \times 1) \times 1](1 \times \Delta)(1 \times kE)\Delta$$

Let $G\colon \mu(D \times 1)(1 \times \Delta) \sim p_1$ rel α ($\alpha\colon X \vee X \to X \times X$, $\alpha(x,*) = x,*$ $\alpha(*,x) = (x,x)$.) $G\colon X \times X \to PX$ yields $\hat{G}\colon PX \times PX \to PPX$ $\hat{G}\colon P[\mu(D \times 1)(1 \times \Delta)] \sim Pp_1 \cdot PE_0\hat{G} = GE_0$ and $PE_\infty\hat{G} = GE_\infty$ imply $\hat{G}(X^{S^1} \times X^{S^1}) \subset PX^{S^1}$ and $\hat{G}(1 \times kE_0)\Delta\colon X^{S^1} \to PX^{S^1}$ is a homotopy $P\mu[PD \times 1](1 \times \Delta)(1 \times kE)\Delta \sim Pp_1(1 \times kE)\Delta = 1$. Consider $PE\colon PX^{S^1} \to PX$.

$$PE\hat{G}(1 \times kE_0)\Delta = GE(1 \times kE)\Delta =$$

$$= G(E \times E)\Delta = G\Delta E$$

As G is rel α it is rel Δ, so $G\Delta E$ is a constant path and

$PE\hat{G}(1 \times kE)\Delta$: $\alpha\gamma \sim 1$ corel E. $\gamma\alpha = (E \times [PD(1 \times kE)\Delta])P\mu(i \times k)T$ and $p_1\gamma\alpha = EP\mu(i \times k)T = \mu(* \times 1)T = p_1$. As α is a homotopy equivalence $\alpha\gamma \sim 1$ implies $\gamma\alpha \sim 1$ or $\gamma\alpha \sim (p_1 \times p_2)\Delta$. As $\gamma\alpha = (p_1 \times \hat{p}_2)$ $\hat{p}_2 \sim p_2$ and $\gamma\alpha \sim 1$ corel p_1.

1.4. H-deviation and H-homotopy equivalence

1.4.1. Definition: Let X,μ, X',μ' be H-spaces. Let $f: X \to X'$ be a map of spaces. H-*deviation of* f is the map

$$HD(f) = HD(f,\mu,\mu') \in [X \wedge X, X']$$

defined by

$$HD(f,\mu,\mu')\Lambda = D_{f\mu,\mu'(f \times f)}$$

If $i: X \vee X \subset X \times X$ then $f\mu i = \mu'(f \times f)i$ implies $D_{f\mu,\mu'(f \times f)}i = *$ and $D_{f\mu,\mu'(f \times f)}$ factors through $\Lambda: X \times X \to X \wedge X$. Moreover, f is an H-map if and only if $f\mu \sim \mu'(f \times f)$ rel $X \vee X$ and this is equivalent to $HD(f,\mu,\mu') = *$. Thus $HD(f,\mu,\mu')$ is the obstruction for f to be a $\mu - \mu'$ H-map: f is a $\mu - \mu'$ H-map if and only if $HD(f,\mu,\mu') = *$.

1.4.2. Proposition: *Let* (X_0, μ_0), (X_1, μ_1) *and* (X_2, μ_2) *be H-spaces,* $f_0: X_0 \to X_1$, $f_1: X_1 \to X_2$ *maps of spaces.*

(a) If f_0 is a μ_0, μ_1 *H-map then* $HD(f_1 f_0) = HD(f_1)(f_0 \wedge f_0)$

(b) If f_1 is a $\mu_1 - \mu_2$ *H-map then* $HD(f_1 f_0) = f_1 HD(f_0)$

Proof:

(a) $\Lambda^*HD(f_1f_0) = D_{f_1f_0\mu_0},\ \mu_2(f_1f_0 \times f_1f_0) =$

$= D_{(f_1\mu_1)(f_0 \times f_0)},\ \mu_2(f_1 \times f_1)(f_0 \times f_0) =$

$= D_{f_1\mu_1},\ \mu_2 f_1 \times f_1{}^{f_0 \times f_0} =$

$= HD(f_1)\Lambda f_0 \times f_0 = HD(f_1)(f_0 \wedge f_0)\Lambda =$

$= \Lambda^*HD(f_1)(f_0 \wedge f_0)$

and as Λ^* is 1-1 (a) follows.

(b) is proven similarly.

Given a space X. The existence of an H-structure for X is not easily detectable and relatively few spaces will admit one (and see Chapter 3 for some necessary conditions). However, if such an H-structure exists it is seldom unique.

Indeed, if $w \in [X \wedge X, X]$ and $\mu: X \times X \to X$ is a multiplication then $\mu_w = w\Lambda + \mu$ is also a multiplication. If $i: X \vee X \subset X \times X$ $i^*\mu_w = i^*w\Lambda + i^*\mu = w\Lambda i + \mu i = F \in [X \vee X, X]$. Obviously $D_{\mu_w,\mu} = w\Lambda$, hence, $HD(1, \mu_w, \mu) = w$. The set of H-structures for X is the set

$$i^{*-1}(F) \subset [X \times X, X], \quad i: X \vee X \subset X \times X.$$

If $\mu \in i^{*-1}(F)$ is fixed then the assignment $\alpha_\mu: i^{*-1}(F) \to [X \wedge X, X]$ given by $\alpha_\mu(\hat{\mu}) = HD(1, \hat{\mu}, \mu)$ is an inverse to the functions $[X \wedge X, X] \to i^{*-1}(F)$ given by $w \to \mu_w$. Hence:

1.4.3. Theorem ([Copeland]$_1$): *If* X, μ *is an H-space then the set of H-structures for* **X** *is in* 1-1 *correspondence with* $[X \wedge X, X]$.

1.4.4. Lemma: *Let* $f: X,\mu \to X',\mu'$ *be an H-map. If* f *is a homotopy equivalence with a homotopy inverse* g *then* g *is an H-map.*

Proof: $* = HD(1,\mu,\mu) = HD(gf, \mu, \mu) = HD(g,\mu',\mu)(f \wedge f)$ and as $(f \wedge f)^*$ is an isomorphism $HD(g,\mu',\mu) = *$.

1.4.5. Definition: Two H-structures μ_1 and μ_2 for X are said to be *H-equivalent* if there exists an *H-homotopy equivalence* $h: X,\mu_1 \to X,\mu_2$, i.e: $h: X \to X$ is a $\mu_1 - \mu_2$ H-map and a homotopy equivalence. This relation is an equivalence relation and the set $i^{*-1}(F)$ modulo this relation is called *the set of essentially different H-structures for* X.

1.4.6. Example: To illustrate how the set of essentially different H-structures can differ from the set of all H-structures consider the following example.

Let $X = K(Z,2n) \times K(Z,4n)$ where $K(G,m)$ is the Eilenberg MacLane space with

$$\pi_n(K(G,m)) = \begin{cases} 0 & m \neq n \\ G & m = n \end{cases}$$

$K(G,m)$ may be considered as an abelian topological group ([Milgram]). Let $\mu_0: X \times X \to X$ be the product multiplication:

$\mu_0[(k_1,k_2)(\overline{k}_1,\overline{k}_2)] = k_1 + \overline{k}_1,\ k_2 + \overline{k}_2$ where $+$ denotes the abelian multiplication in $K(Z,m)$, $m = 2n, 4n$.

Now, $[X \wedge X,X] = H^{2n}(X \wedge X,Z) \oplus H^{4n}(X \wedge X,Z)$. As X is $2n-1$ connected $X \wedge X$ is $4n-1$ connected and $H^{4n}(X \wedge X,Z) = Z$, hence $[X \wedge X,X] \approx Z$. The generator $u \in [X \wedge X,X] = Z$ is given by

$$u[(k_1,k_2),\ (\overline{k}_1,\overline{k}_2)] = *,\quad w(k_1,\overline{k}_1),$$

$w \in [K(Z,2n) \wedge K(Z,2n),\ K(Z,4n)] = H^{4n}(K(Z,2n) \wedge K(Z,2n),\ Z) = Z$ is a generator: $w = \iota_{2n} \otimes \iota_{2n}$, $\iota_{2n} \in H^{2n}(K(Z,2n),Z)$ - the fundamental class.

If $\lambda \in Z$ the element in $i^{*-1}(F)$ corresponding to $\lambda u \in [X \wedge X, X]$ is the multiplication μ_λ given by

$$\mu_\lambda[(k_1,k_2),(\bar{k}_1,\bar{k}_2)] = k_1 + \bar{k}_1,\ \lambda w(k_1,\bar{k}_1) + k_2 + \bar{k}_2$$

Let $g\colon K(Z,2n) \to K(Z,4n)$ be given by $g(x) = w(x,x)$. For any $\nu \in Z$ let $h_\nu\colon X \to X$ be given by $h_\nu(k_1,k_2) = k_1,\ \nu g(k_1) + k_2$. As $h_{-\nu}h_\nu = 1$ h_ν are homeomorphisms.

It is obvious that X,μ_λ and $X,h_\nu\mu_\lambda(h_{-\nu} \times h_{-\nu})$ are H-homotopy equivalent. Now,

$$\begin{aligned} h_\nu\mu_\lambda(h_{-\nu} \times h_{-\nu})[(k_1,k_2),\ (\bar{k}_1,\bar{k}_2)] &= \\ = k_1 + \bar{k}_1,\ \nu(g(k_1 + \bar{k}_1) - g(k_1) - g(\bar{k}_1)) &+ \\ + \lambda w(k_1,\bar{k}_1) + k_2 + \bar{k}_2 & \end{aligned}$$

To investigate the map $\hat{g}\colon K(Z,2n) \wedge K(Z,2n) \to K(Z,4n)$

$$\hat{g}(k_1,\bar{k}_1) = g(k_1 + \bar{k}_1) - g(k_1) - g(\bar{k}_1)$$

one should study $H^{4n}(\hat{g},Z)$.

$$\hat{g} = gm - gp_1 - gp_2 \quad (m \text{ the addition, } p_i \text{ - the projections})$$

$H^{4n}(g,Z)\ \iota_{4n} = \iota_{2n}^2$ $H^*(m,Z)\iota_{2n} = \iota_{2n} \otimes 1 + 1 \otimes \iota_{2n}$ $H^*(p_1,Z)\ \iota_{2n} = \iota_{2n} \otimes 1$ $H^*(p_2,Z)\iota_{2n} = 1 \otimes \iota_{2n}$ and recall that for every f $H^*(f,Z)$ is a ring homomorphisms, thus

$$H^{4n}(\hat{g},Z)\iota_{4n} = H^*(m,Z)\iota_{2n}^2 - H^*(p_1,Z)\iota_{2n}^2 - H^*(p_2,Z)\iota_{2n}^2 =$$
$$= (\iota_{2n} \otimes 1+1 \otimes \iota_{2n})^2 - \iota_{2n}^2 \otimes 1-1 \otimes \iota_{2n}^2 =$$
$$= 2(\iota_{2n} \otimes \iota_{2n}) = 2w$$

It follows that $\hat{g} \sim 2w$ and $h_\nu \mu_\lambda (h_{-\nu} \times h_{-\nu}) = \mu_{\lambda+2\nu}$. Consequently, there are at most two essentially different H-structures for X: μ_0 and μ_1. (Actually, using the Hopf algebra structures of $H^*(X,Z_2)$ induced by μ_0 and μ_1 as discussed in Chapter 3 one can see that $H^*(\mu_0,Z_2) \neq H^*(\mu_1,Z_2)$: Let $0 \neq x \in \text{im}\ H^{2n}(K(Z,2n),Z_2) \to H^{2n}(X,Z_2)$, $0 \neq y \in \text{im}\ H^{4n}(K(Z,4n),Z_2) \to H^{4n}(X,Z_2)$. Then

$$H^*(\mu_0,Z_2)y = y \otimes 1 + 1 \otimes y$$
$$H^*(\mu_1,Z_2)y = y \otimes 1 + 1 \otimes y + x \otimes x$$

Hence, μ_0 and μ_1 induce non isomorphic Hopf algebra structures on $H^*(X,Z_2)$ and are essentially different).

For further study of the enumeration of H-structures and essentially different H-structures see [Arkowitz and Curjel]$_1$ and [Curjel].

1.5. Change of H-structures and H-maps.

It is obvious that the possession of an H-structure is a homotopy invariant. Moreover, certain homotopy constructions preserve this property (see Chapter 2): Starting with H-spaces and H-maps the outcome of the construction is another H-space. Quite often the specific multiplications of the given H-spaces are not important. Thus, one may ask: Given spaces X and X' and a map $h: X \to X'$, are there multiplicactions μ and μ' so that $h: X,\mu \to X',\mu'$ is a $\mu - \mu'$ H-map?

A useful test is given by the following:

<u>1.5.1. Proposition</u>: *Let* (X,μ), (X',μ') *be H-spaces and* $h\colon X \to X'$ *a map of spaces.*

(a) X' *admits a multiplication* $\hat{\mu}'$ *so that* h *is a* $\mu - \hat{\mu}'$ *H-map if and only if*

$$HD(h,\mu,\mu') \in \operatorname{im}([X' \wedge X',X'] \xrightarrow{(h\wedge h)^*} [X \wedge X,X'])$$

(b) Suppose X',μ' *is HA and either* X *is finite dimensional or* $\pi_n(X') = 0$ *for* n *sufficiently large. Then* X *admits a multiplication* $\hat{\mu}$ *so that* h *is a* $\hat{\mu} - \mu'$ *H-map if and only if*

$$-HD(h,\mu,\mu') \in \operatorname{im}([X \wedge X,X] \xrightarrow{h_*} [X \wedge X,X'])$$

<u>Proof</u>:

(a) Suppose $HD(h,\mu,\mu') \sim w(h \wedge h)$, $w \in [X' \wedge X',X']$.
Let $\hat{\mu}' = \mu'_w = w\Lambda + \mu'$. (or $\hat{\mu}' = \mu'(w\Lambda \times \mu')\Delta_{X' \times X'}$). Then:

$$\begin{aligned}\hat{\mu}'(h \times h) &= \mu'(w\Lambda \times \mu')\Delta_{X' \times X'}(h \times h) = \\ &= \mu'[w(h \wedge h)\Lambda \times \mu'(h \times h)]\Delta_{X \times X} = \\ &= \mu'[HD(h,\mu,\mu')\Lambda \times \mu'(h \times h)]\Delta_{X \times X} = h\mu\end{aligned}$$

Conversely, if $\hat{\mu}' = \mu'_w(w \in [X' \wedge X',X'])$ is an H-structure for X' so that h is a $\mu - \hat{\mu}'$ H-map then

$$h\mu = \hat{\mu}'(h \times h) = \mu'[w(h \wedge h)\Lambda \times \mu'(h \times h)]\Delta_{X \times X}$$

and $HD(h,\mu,\mu') = w(h \wedge h)$.

(b) Suppose $\hat{\mu}: X \times X \to X$ exists so that h is a $\hat{\mu} - \mu'$ H-map. $\mu = \hat{\mu}_w = w\Lambda + \hat{\mu}$, $w \in [X \wedge X, X]$ and $+$ denotes the operation in $[\ ,X]$ induced by $\hat{\mu}$.

$$h\mu = h\hat{\mu}(w\Lambda \times \hat{\mu})\Delta_{X \times X} =$$

$$= \mu'(hw\Lambda \times h\hat{\mu})\Delta_{X \times X} =$$

$$= \mu'(hw\Lambda \times \mu'(h \times h))\Delta_{X \times X} \quad \text{and}$$

$$hw = HD(h,\mu,\mu')$$

If $cw \in [X \wedge X, X]$ is given by $cw = D_{*,w}$ (and D is the difference taken with respect to $\hat{\mu}$) then

$$hcw = hD_{*,w} = D_{*,hw} = -hw = -HD(h,\mu,\mu').$$

The converse is slightly more complex and here the finiteness condition in the hypothesis is needed:

Suppose $-HD(h,\mu,\mu') = hv$, $v \in [X \wedge X, X]$. Denote $w = HD(h,\mu,\mu')$. Consider $\mu_1 = \mu_v = v\Lambda + \mu$ (and here $+$ stands for the operation in $[\ ,X]$ induced by μ). Then:

$$h\mu_1 = h_*(v\Lambda + \mu) = w\Lambda(v\Lambda \times \mu)\Delta_{X \times X} + hv\Lambda + h\mu =$$

$$= w\Lambda(v\Lambda \times \mu)\Delta_{X \times X} + hv\Lambda + w\Lambda + \mu'(h \times h)$$

As $hv + w = 0$ one gets

$$h\mu_1 = w\Lambda(v\Lambda \times \mu)\Delta_{X \times X} + \mu'(h \times h)$$

and

$$HD(h,\mu_1,\mu')\Lambda = w\Lambda(v\Lambda \times \mu)\Delta_{X \times X} = w[(v\Lambda) \wedge \mu]\Lambda\Delta_{X \times X}$$

Let $\overline{\Delta}_{X \times X}: X \times X \to (X \times X) \wedge (X \times X)$ be induced by the diagonal $\Delta_{X \times X}$ $\overline{\Delta}_{X \times X} = \Lambda\Delta_{X \times X}$. There exists $\alpha_1: X \wedge X \to (X \wedge X) \wedge (X \times X)$ so that $\alpha_1\Lambda = (\Lambda \wedge 1)\overline{\Delta}_{X \times X}$ hence,

$$((v\Lambda) \wedge \mu)\overline{\Delta}_{X \times X} = (v \wedge \mu)(\Lambda \wedge 1)\overline{\Delta}_{X \times X} =$$
$$= (v \wedge \mu)\alpha_1\Lambda \quad \text{and}$$

$w_1 = HD(h,\mu_1,\mu') = w(v \wedge \mu)\alpha_1 = -h(v(v \wedge \mu)\alpha_1)$. Put
$v_1 = v(v \wedge \mu)\alpha_1 = v(v \wedge 1)\overline{\alpha}_1(1 \wedge \mu)\alpha_1$.

Suppose inductively that μ_n, v_n were constructed so that $\mu_n: X \times X \to X$ is a multiplication with $-w_n = -HD(h,\mu_n,\mu') = hv_n$ and v_n factors as $X \wedge X \xrightarrow{\overline{\alpha}_n} \Lambda^{n+2}X \xrightarrow{\hat{v}_n} X$ where $\Lambda^{n+2}X$ is the n+2 fold smash product of X.

In the computation leading from μ, v to v_1 replace μ by μ_n, v by v_n and $\mu_{n+1} = (\mu_n)_{v_n} = v_n\Lambda + \mu_n$ (where here + denotes the addition in $[\ ,X]$ induced by μ_n). Then one obtains
$-w_{n+1} = -HD(h, \mu_{n+1}, \mu') = hv_n(v_n \wedge 1)(1 \wedge \mu_n)\alpha_1$.
$v_{n+1} = v_n(v_n \wedge 1)(1 \wedge \mu_n)\alpha_1$ can be factored as

$$\begin{array}{ccccc} X \wedge X & \xrightarrow{(1 \wedge \mu_n)\alpha_1 = \gamma_n} & X \wedge X \wedge X & \xrightarrow{\overline{\alpha}_n \wedge 1} & (\Lambda^{n+1}X) \wedge X \\ & & \searrow^{v_n \wedge 1} & & \swarrow_{\hat{v}_n \wedge 1} \\ & & & X \wedge X \xrightarrow{v_n} X & \end{array}$$

Put $\overline{\alpha}_{n+1} = (\overline{\alpha}_n \wedge 1)\gamma_n$, $\hat{v}_{n+1} = v_n(\hat{v}_n \wedge 1)$ to complete the inductive step:

$$v_{n+1} = \hat{v}_{n+1}\overline{\alpha}_{n+1}$$

As $\Lambda^n X$ is at least $n-1$ connected if $\dim X = k$

$$[X \wedge X,\ \Lambda^{2k+1}X] = * \text{ and } \bar{\alpha}_{2k-1} = * = v_{2k-1}$$

and $-hv_{2k-1} = HD(h,\ \mu_{2k-1},\ \mu') = *$. If $\pi_n(X') = 0$ for $n > k$ then $[\Lambda^{k+1}X,\ X'] = *$, $h\hat{v}_{k-1} \sim * \sim hv_{k-1}$ and $HD(h,\ \mu_{k-1},\ \mu') = *$.

Chapter II

Homotopy Properties of H-spaces

Introduction

In this chapter some deeper homotopy properties of H-spaces are studied: fibrations, Postnikov (and Moore) approximations and various obstructions in the theory of homotopy-associative and homotopy-commutative H-spaces. The notions of actions and principal fibrations in the context of homotopy theory and H-spaces are considered, as well as the notions of homotopy nilpotency and homotopy solvability.

As the computations become more complex and the amount of homotopy theory used is wider, it is naturally expected that the reader's experience should be somewhat greater than that required in Chapter I.

2.1. H-spaces and fibrations.

Let $f_i, F_i\colon X_i, \mu_i \to Y, \mu$ $i = 1,2$ be H-maps (in $\mathcal{H}_{fF}$). Let $p_i\colon W_{f_1,f_2} \to X_i$ $i = 1,2$ be the homotopy pullback of f_1 and f_2:

$$W_{f_1,f_2} = \{x_1, \overline{\gamma}, x_2 \in X_1 \times PY \times X_2 \mid \varepsilon_0(\overline{\gamma}) = f_1x_1, \varepsilon_\infty(\overline{\gamma}) = f_2x_2\}$$

$$p_i(x_1, \overline{\gamma}, x_2) = x_i \in X_i$$

One can define in W_{f_1,f_2} the following product:

$$\hat{\mu}[(x_1,\overline{\gamma},x_2), (x_1',\overline{\gamma}',x_2')] = \tag{2.1.1}$$

$$\mu_1(x_1,x_1'),\ F_1(x_1,x_1') + P\mu(\overline{\gamma},\overline{\gamma}') - F_2(x_2,x_2'),\ \mu_2(x_2,x_2')$$

$\hat{\mu}$ is indeed a multiplication as

$$\hat{\mu}[(*,*,*),\ (x_1',\overline{\gamma}',x_2')] = \mu(*,x_1'),\ F_1(*,x_1') + P\mu(*,\overline{\gamma}') -$$
$$- F_2(*,x_2'),\ \mu_2(*,x_2') = x_1',\overline{\gamma}',x_2'$$

and similarly for $\hat{\mu}[(x_1,\overline{\gamma},x_2),\ (*,*,*)]$.

p_i are obviously H-maps. Hence one has

2.1.2. Theorem (Moore-Stasheff, see [Stasheff]$_1$ and [Stasheff]$_5$):

If $f_i: X_i,\mu_i \to Y,\mu$ *are H-maps* $i = 1,2$, *then the pullback* W_{f_1,f_2} *of* f_1 *and* f_2 *and* $p_i: W_{f_1,f_2} \to X_i$ *admit H-structures.*

A slightly more general pull back theorem is given by:

2.1.3. Proposition: *Let* X_i,μ_i $i = 1,2$ *be H-spaces and let* Y *be a space. Suppose there exist maps* $\eta_1: Y \times X_1 \to Y$, $\eta_2: X_2 \times Y \to Y$ *so that* $\eta_1(y,*) = y$, $\eta_2(*,y) = y$. *Let* $h_i: X_i \to Y$ *be given by* $h_1(x_1) = \eta(*,x_1)$, $h_2(x_2) = \eta_2(x_2,*)$. *Suppose the following three diagrams commute*

(1)
$$\begin{array}{ccc} X_1 \times X_1 & \xrightarrow{\mu_1} & X_1 \\ {\scriptstyle h_1 \times 1}\downarrow & & \downarrow{\scriptstyle h_1} \\ Y \times X_1 & \xrightarrow{\eta_1} & Y \end{array}$$

(2)
$$\begin{array}{ccc} X_2 \times X_2 & \xrightarrow{\mu_2} & X_2 \\ {\scriptstyle 1 \times h_2}\downarrow & & \downarrow{\scriptstyle h_2} \\ X_2 \times Y & \xrightarrow{\eta_2} & Y \end{array}$$

(3)
$$\begin{array}{ccc} X_2 \times X_1 & \xrightarrow{h_2 \times 1} & Y \times X_1 \\ {\scriptstyle 1 \times h_1}\downarrow & & \downarrow{\scriptstyle \eta_1} \\ X_2 \times Y & \xrightarrow{\eta_2} & Y \end{array}$$

Then the pullback W_{h_1,h_2} *and the maps* $W_{h_1,h_2} \to X_i$ *admit H-structures.* [Note that if Y admits an H-structure μ and $h_i\colon X_i,\mu_i \to Y,\mu$ are H-maps, then $\eta_1 = \mu(1 \times h_1)$, $\eta_2 = \mu(h_2 \times 1)$ satisfy the hypothesis of this theorem.]

Proof: We first show that the homotopies of diagrams (1), (2) and (3) can be chosen to be rel $X_i \vee X_j$. (On $X_i \vee X_j$, the homotopic maps actually agree. If Y is an H-space, this follows from 1.1.5.)

Let $\hat{F}_1\colon X_1 \times X_1 \to PY$ be a homotopy $h_1\mu_1 \sim \eta_1(h_1 \times 1)$. Define $\tilde{F}_1\colon X_1 \times X_1 \to PY$ by

$$\tilde{F}_1(x,x') = -\hat{F}_1(*,\mu_1(x,x')) + \hat{F}_1(x,x') -$$
$$- P\eta_1(\hat{F}_1(x,*),kx') + P\eta_1(\hat{F}_1(*,x),kx')$$

One can then see that $\tilde{F}_1$ is well-defined and is a homotopy $h_1\mu_1 \sim \eta_1(h_1 \times 1)$. Now,

$$\tilde{F}_1(*,x') = -\hat{F}_1(*,x') + \hat{F}_1(*,x')$$
$$\tilde{F}_1(x,*) = -\hat{F}_1(*,x) + \hat{F}_1(x,*) - \hat{F}_1(x,*) + \hat{F}_1(*,x)$$

Hence $\tilde{F}_1 | X_1 \vee X_1 \sim kh_1F$ corel E_0, E_∞ and by standard homotopy theory $\tilde{F}_1 \sim F_1$ corel E_0, E_∞ and $F_1 | X_1 \vee X_1 = kh_1F$. Hence $F_1\colon h_1\mu_1 \sim \eta_1(h_1 \times 1)$ rel $X_1 \vee X_1$.

Similarly, one has $F_2\colon h_2\mu_2 \sim \eta_2(1 \times h_2)$ rel $X_2 \vee X_2$. Let $\hat{F}_3\colon X_2 \times X_1 \to PY$ be a homotopy $\eta(h_2 \times 1) \sim \eta_2(1 \times h_1)$. Let

$$\tilde{F}_3(x_2,x_1) = -P\eta_1(\hat{F}_3(x_2,*),kx_1) + \hat{F}_3(x_2,x_1) - P\eta_2(kx_2,\hat{F}_3(*,x_1))$$

and proceed similarly.

If F_i are homotopies rel $X_i \vee X_j$ corresponding to diagram (i) $i = 1,2,3$ define $\hat{\mu}\colon W_{h_1,h_2} \times W_{h_1,h_2} \to W_{h_1,h_2}$ by

$$\begin{aligned}\hat{\mu}[(x_1,\bar{\gamma},x_2),\ (x_1',\bar{\gamma}',x_2')] = \mu_1(x_1,x_1'),\ & F_1(x_1,x_1') + \\ & + P\eta_1(\bar{\gamma},kx_1') + F_3(x_2,x_1') + \\ & + P\eta_2(kx_2,\bar{\gamma}') - F_2(x_2,x_2'),\ \mu_2(x_2,x_2')\end{aligned}$$

<u>2.1.4. Corollary</u>: *Let* X, μ *be an* H-*space,* $\eta\colon X \times Y \to Y$ *a map satisfying* $\eta(*,y) = y$ *and putting* $h(x) = \eta(x,*)$, *supposing that* $h\mu \sim \eta(1 \times h)$. *Then the fiber of* h *admits an* H-*structure* $\hat{\mu}$ *so that* fiber $h \to X$ *is a* $\hat{\mu}$-μ H-*map.*

<u>Proof</u>: Put $X_1 = *$, $\eta_1 = 1$, $X_2 = X$, $\eta_2 = \eta$ and apply 2.1.3.

<u>2.2. H-liftings</u>

Any lifting problem in the category of H-spaces can be dealt with in two stages: First, one seeks a solution to the ordinary lifting problem, and then one checks if one of the solutions can be chosen to serve as an H-map. In the context of the theory of H-spaces, it is therefore natural to investigate the obstruction for the existence of an H-lifting, provided an ordinary lifting exists. More precisely, given H-spaces and H-maps:

$$(f_0,F_0)\colon\ X_0,\mu_0 \longrightarrow X_1,\mu_1$$
$$(g,G)\colon\ Y,\mu \longrightarrow X_1,\mu_1$$

If $f\colon X_0 \to Y$ lifts f_0 (as a map) $gf \sim f_0$ then $gHD(f,\mu_0,\mu) = HD(f_0,\mu_0,\mu_1) = *$. Hence $HD(f,\mu_0,\mu)\colon X_0 \wedge X_0 \longrightarrow Y$ lifts to a map $w\colon X_0 \wedge X_0 \longrightarrow V_g$ = Fiber g, $j\colon V_g \to Y$, $HD(f,\mu_0,\mu) = jw$. $jw = *$ implies that w lifts to $w\colon X_0 \wedge X_0 \longrightarrow$ fiber $j \approx \Omega X_1$ and the obstruction for f to be an H-map is

$$[w] \in [X_0 \wedge X_0,\ V_g]/\mathrm{im}\{[X_0 \wedge X_0, \Omega X_1] \longrightarrow [X_0 \wedge X_0,\ V_g]\}.$$

This set in its simplest form could be considered as the set $[X_0 \wedge X_0,\ V_g]$, with the denominator subset identified to a point. In general, $[X_0 \wedge X_0,\ \Omega X_1]$ is a group acting on $[X_0 \wedge X_0,\ V_g]$ and $[w]$ could be considered as an element of the orbit set.

Without further assumptions, it is hard to measure the effect that the change of the lifting f will have on $[w]$. Suppose now $g\colon Y \to X_1$ is the fiber $W_{f_1,*}$ of an H-map $f_1,F_1\colon X_1,\mu_1 \to X_2,\mu_2$ and μ is the 2.1.1 multiplication induced by μ_1,μ_2 and F_1.

The liftings $f\colon X_0 \to Y$ are in 1-1 correspondence with homotopies $\xi\colon * \sim f_1 f_0$, $\xi\colon X_0 \to LX_2$ via the assignment $\xi \to f_\xi$ $f_\xi(x) = f_0(x), \xi(x)$ $f_\xi(x) \in Y$. $gf_\xi = f_0$ and as g is multiplicative, one may assume that $G = kg\mu$. The H-lifting problem (modulo the ordinary lifting problem) is that of finding $F\colon X_0 \times X_0 \to PY$ so that $f_\xi = f,F$ is an H-map. Such an F exists if, and only if, the following obstruction $\widetilde{D}$ vanishes.

$$\widetilde{D}\colon X_0 \times X_0 \to \Omega X_2$$
$$\widetilde{D} = \widetilde{D}(F_0,F_1,\xi) = \xi\mu_0 + Pf_1F_0 + F_1(f_0 \times f_0) - P\mu_2(\xi \times \xi)$$

Further, as $\widetilde{D}|X_0 \vee X_0 = \xi F - PF(\xi \vee \xi) = \xi F - \xi F \sim *$, one can find $D \sim \widetilde{D}$ so that $D|X \vee X = *$ and if $D \sim *$ by 1.1.5, $D \sim *$ rel $X \vee X$. Alternatively, one can always assume that if $\widetilde{D} \sim *$, then this homotopy extends the obvious homotopy $\widetilde{D}|X \vee X \sim *$. It follows that one can replace the obstruction $\widetilde{D}$ by $D: X_0 \wedge X_0 \longrightarrow \Omega X_2$, $D = D(\xi, F_0, F_1)$.

D actually depends on μ_0, μ_2, f_0, f_1 as well as on ξ, F_0, F_1 but it is only the last three that we allow to vary, in order to annihilate the obstruction D.

<u>2.2.1. Proposition</u>: *Let* X_i, μ_i $i = 0,1,2$ *be H-spaces* $f_i, F_i: X_i, \mu_i \to X_{i+1}, \mu_{i+1}$ $i = 0,1$ *H-maps* $\xi: X_0 \to LX_2$ *a homotopy* $* \sim f_1 f_0$.

Given maps

$$w_0: X_0 \wedge X_0 \longrightarrow \Omega X_1$$

$$w_1: X_1 \wedge X_1 \longrightarrow \Omega X_2$$

$$w : X_0 \longrightarrow \Omega X_2$$

Put

$$\hat{F}_0 = P\mu_1(w_0\Lambda \times F_1)\Delta_{X_0 \times X_0}$$

$$\hat{F}_1 = P\mu_2(w_1\Lambda \times F_1)\Delta_{X_1 \times X_1}$$

$$\hat{\xi} = P\mu_2(w \times \xi)\Delta_{X_0}$$

Then:

(a) $f_i, \hat{F}_i$ *are H-maps and*

$$D(\hat{F}_0, \hat{F}_1, \hat{\xi}) = D(F_0, F_1, \xi) + \Omega f_1 w_0 + w_1(f_0 \wedge f_0) + HD(w, \mu_0, \Omega\mu_2)$$

(b) Let $Y = W_{*,f_1}$ *be the fiber of* f_1 *with the* 2.1.1 *multiplication* $\mu = \mu(\mu_1, \mu_2, F_1)$. *Let* $f_\xi = f: X_0 \to Y$ *be the lifting of* f_0 *induced by* ξ: $f_\xi(x) = f(x) = f_0(x), \xi(x)$. *Then* $HD(f, \mu_0, \mu) = jD(F_0, F_1, \xi)$ *where* $j: \Omega X_2 \to Y$

is the inclusion.

Proof: (a) $f_i, \hat{F}_i$ are obviously H-maps. Put $\hat{w}_i = w_i\Lambda$. By 1.2.4 and 1.2.5:

$$\hat{F}_i \sim P\mu_{i+1}(\hat{w}_i \times kE_0F_i)\Delta_{X_i \times X_i} + F_i, \quad \hat{\xi} \sim w+\xi \quad \text{corel } E_0, E_\infty$$

$$\begin{aligned}\widetilde{D}(\hat{F}_0,\hat{F}_1,\hat{\xi}) &= \hat{\xi}\mu_0 + Pf_1\hat{F}_0 + \hat{F}_1(f_0 \times f_0) - P\mu_2(\hat{\xi} \times \hat{\xi}) = \\ &= w\mu_0 + \xi\mu_0 + Pf_1P\mu_1(\hat{w}_0 \times kE_0F_0)\Delta_{X_0 \times X_0} + \\ &+ Pf_1F_0 + P\mu_2(\hat{w}_1 \times kE_0F_1)\Delta_{X_1 \times X_1}(f_0 \times f_0) + \\ &+ F_1(f_0 \times f_0) - P\mu_2(\xi \times \xi) - P\mu_2(w \times w).\end{aligned}$$

Now, $Pf_1P\mu_1 \sim P\mu_2(Pf_1 \times Pf_1)$ and on $\Omega X \times PX$ this homotopy is corel E_0, E_∞. Hence

$$\begin{aligned}\widetilde{D} &= w\mu_0 + \xi\mu_0 + P\mu_2(\Omega f_1\hat{w}_0 \times kE_0Pf_1F_0)\Delta_{X_0 \times X_0} + Pf_1F_0 \\ &+ P\mu_2[\hat{w}_1(f_0 \times f_0) \times kE_0F_1(f_0 \times f_0)]\Delta_{X_0 \times X_0} + F_1(f_0 \times f_0) - \\ &- P\mu_2(\xi \times \xi) - P\mu_2(w \times w)\end{aligned}$$

Now again by 1.2.4 and 1.2.5 if $\gamma_1,\gamma_2 \in P_2(X)$ $\lambda \in \Omega X$ $X,\hat{\mu}$ - an H-space then the maps

$$\gamma_1,\lambda,\gamma_2 \to \gamma_1 + P\hat{\mu}(\lambda, kE_0\gamma_2) + \gamma_2 \quad \text{and}$$

$$\gamma_1,\lambda,\gamma_2 \to \gamma_1 + \gamma_2 + P\hat{\mu}(\lambda, kE_\infty\gamma_2)$$

are homotopic and as ΩX_2, Add is homotopy commutative

$$\begin{aligned}\widetilde{D} &\sim \xi\mu_0 + Pf_1F_0 + F_1(f_0 \times f_0) - P\mu_2(\xi \times \xi) + \Omega f_1\hat{w}_0 + \\ &+ \hat{w}_1f_0 \times f_0 + w\mu_0 - \Omega\mu_2(w \times w) = \\ &= \widetilde{D}(\xi,F_0,F_1) + [\Omega f_1w_0 + w_1f_0 \wedge f_0 + HD(w,\mu_0,\Omega\mu_2)]\Lambda\end{aligned}$$

One can see that all homotopies are constant on $X_0 \vee X_0$ hence one may add the homotopy $\tilde{D}(\xi, F_0, F_1) \sim D(\xi, F_0, F_1)\Lambda$ to complete the proof of (a).

(b) $\mu(j \times 1): \Omega X_2 \times Y \to Y$ is given by $\lambda, (x, \overline{Y}) \to x, \lambda + \overline{Y}$ and coincides with the action. Observe now that the following homotopies all remain in Y.

$$\mu(f_\xi \times f_\xi) = \{\mu_1(f_0 \times f_0) \times [P\mu_2(\xi \times \xi) - F_1(f_0 \times f_0)]\}\Delta_{X_0 \times X_0}$$

$$\begin{aligned} f_\xi \mu_0 &= (f_0\mu_0 \times \xi\mu_0)\Delta_{X_0 \times X_0} \sim \{[\mu_1(f_0 \times f_0)] \times [\xi\mu_0 + pf_1F_0]\}\Delta_{X_0 \times X_0} \\ &\sim \{[\mu_1(f_0 \times f_0)] \times [(\xi\mu_0 + Pf_1F_0 + F_1(f_0 \times f_0) - P\mu_2(\xi \times \xi)) + \\ &\quad + (P\mu_2(\xi \times \xi) - F_1(f_0 \times f_0))]\}\Delta_{X_0 \times X_0} = \\ &= \mu(j\tilde{D}(\xi, F_0, F_1) \times \mu(f_\xi \times f_\xi))\Delta_{X_0 \times X_0} \end{aligned}$$

2.2.1 (b) follows by replacing $\tilde{D}$ by $D\Lambda$.

<u>2.2.2. Corollary</u>: *Let* $f_0: X_0, \mu_0 \to X_1, \mu_1$ *and* $f_1: X_1, \mu_1 \to X_2, \mu_2$. *Given H-structures* F_i *to* f_i $i = 0,1$ *and a homotopy* $\xi: * \sim f_1 f_0$. *Suppose*

$$D(\xi, F_0, F_1) \in \mathrm{im}(f_0 \wedge f_0)^* + \mathrm{im}(\Omega f_1)_* + \mathrm{im}HD$$

$$(f_0 \wedge f_0)^*: \ [X_1 \wedge X_1, \Omega X_2] \to [X_0 \wedge X_0, \Omega X_2]$$

$$(\Omega f_1)_*: [X_0 \wedge X_0, \Omega X_1] \to [X_0 \wedge X_0, \Omega X_2]$$

$$HD = HD(\ , \mu_0, \Omega\mu_2): [X_0, \Omega X_2] \to [X_0 \wedge X_0, \Omega X_2]$$

Then the fiber Y *of* f_1 *admits an H-structure* μ *over* X_1, μ_1 *(i.e.,* $Y \to X_1$ *is a* $\mu - \mu_1$ *H-map) so that* f_0 *lifts to a* $\mu_0 - \mu$ *H-map.*

2.3. Postnikov systems

Given a space X. *A Postnikov system* for X is a system $\{X_n, h_n, h_{n,n-1}, k_n\}$ of spaces and maps so that

(P1) $h_n: X \to X_n$ satisfies: $\pi_m(h_n)$ is an isomorphism for $m \leq n$. $\pi_m(X_n) = 0$ for $m > n$.

(P2) $h_{n,n-1}: X_n \to X_{n-1}$ is a fibration induced by a map $k_n: X_{n-1} \to K(\pi_n(X), n+1)$ called the n-th k-invariant of X.

(P3) $h_{n,n-1}h_n \sim h_{n-1}$

In our context of spaces every space admits a Postnikov system. We denote $X_n = Ht_n(X)$ and refer to $h_n: X \to Ht_n(X)$ as the homotopy approximation of X in $\dim \leq n$.

If Y is a space with $\pi_m(Y) = 0$ for $m > n$ then any map $f: X \to Y$ factors uniquely through $Ht_n(X)$, $f \sim Ht_n(f)h_n$, $Ht_n(f): Ht_n(X) \to Y$.

A cellular structure for $Ht_n(X)$ can be given by

$$Ht_n(X) = X \cup (\vee_\alpha e_\alpha^{n+2}) \cup \ldots$$

hence, if $C(h_n)$ is the mapping cone of h_n for $m \leq n+1$ $H_m(C(h_n),G) = 0 = H^m(C(h_n),G)$ for every group G of coefficients. It follows that

$$H_m(h_n,G): \ H_m(X,G) \to H_m(Ht_n(X),G)$$

is an isomorphism for $m \leq n$ and an epimorphism for $m = n+1$. Similarly

$$H^m(h_n,G): \ H^m(Ht_n(X),G) \to H^m(X,G)$$

is an isomorphism for $m \leq n$ and a monomorphism for $m = n+1$.

If $h_n: X \to Ht_n(X)$ is a homotopy approximation in $\dim \leq n$ the fiber of h_n, $j_{n+1}: X^{n+1} \to X$ is called the *n-connective fibering of* X.

Given a map $f: X \to Y$ a Postnikov decomposition of f is a system $\{Y_n, f_n, g_n, h_n, k_n\}$ of spaces and maps

$$f_n: X \to Y_n, \quad g_n: Y_n \to Y = Y_0, \quad h_n: Y_n \to Y_{n-1}$$

satisfying: $g_n f_n \sim f$, $h_n f_n \sim f_{n-1}$, $g_n h_{n+1} \sim g_{n+1}$, $\pi_m(f_n)$ is an isomorphism for $m \leq n$ and an epimorphism for $m = n+1$, $\pi_m(g_n)$ is an isomorphism for $m > n+1$ and a monomorphism for $m = n+1$. $h_n: Y_n \to Y_{n-1}$ is the fiber of a map $k_n: Y_{n-1} \to K(\pi_n(\text{fiber } f), n+1)$.

The Postnikov system of a space coincides with the Postnikov decomposition of $X \to *$.

In case $\pi_m(\text{fiber } f)$ are all finite groups, it is sometimes more convenient to further decompose $h_n: Y_n \to Y_{n-1}$ to a sequence of fibrations

$$Y_n = Y_{n,m} \to Y_{n,m-1} \to \cdots \to Y_{n,k} \xrightarrow{h_{n,k}} Y_{n,k-1} \to \cdots Y_{n,0} = Y_{n-1}$$

where $h_{n,j}: Y_{n,j} \to Y_{n,j-1}$ is the fiber of $k_{n,j}: Y_{n,j-1} \to K(Z_{p_j}, n)$ $\{p_j\}$ - primes, $h_n \sim h_{n,1} h_{n,2} \cdots h_{n,j} \cdots h_{n,m}$. This factorization corresponds to a cofiltration of $\pi_n(\text{fiber } f)$:

$\pi_n(\text{fiber } f) = G_m \xrightarrow{\alpha_m} G_{m-1} \xrightarrow{\alpha_{m-1}} \cdots \longrightarrow G_1 \longrightarrow G_0 = 0 \quad \ker \alpha_k = Z_{p_k}$.
This is done by

$$\begin{array}{ccccccccc}
Y_n = Y_{n,m} & \longrightarrow & Y_{n,m-1} & \longrightarrow \cdots \longrightarrow & Y_{n,k} & \xrightarrow{h_{n,k}} & Y_{n,k-1} & \to \cdots \to & Y_{n,0} = Y_{n-1} \\
\downarrow & & \downarrow & & \downarrow & & \downarrow & & \downarrow \\
Y_{n-1} & \xrightarrow{1} & Y_{n-1} & \longrightarrow \cdots \longrightarrow & Y_{n-1} & \xrightarrow{1} & Y_{n-1} & \to \cdots \to & Y_{n-1} \\
\downarrow{\scriptstyle k_n} & & \downarrow & & \downarrow & & \downarrow & & \downarrow \\
K(G_m, n) & \xrightarrow{(\alpha_m)} & K(G_{m-1}, n) & \to \cdots \to & K(G_k, n) & \xrightarrow{(\alpha_k)} & K(G_{k-1}, n) & \to \cdots & *
\end{array}$$

where all vertical triads are fibrations.

Such a system will be referred to as a *primitive Postnikov system.*

With our notion of spaces every map admits a Postnikov and a primitive Postnikov approximations. For details and proofs, see e.g. [Hu] p. 155 or [Spanier] p. 440.

An "Eckmann-Hilton dual" of the Postnikov approximation and decomposition is the *Moore (or homology) approximation and decomposition:*

If $f\colon X \to Y$ then the Moore decomposition of f is a sequence $Y^{(n)}$, $f^{(n)}$, $g^{(n)}$, $k^{(n)}$ of spaces

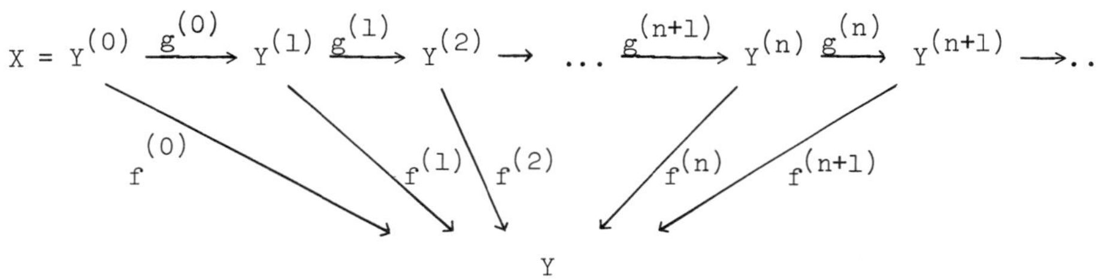

$g^{(n)}\colon Y^{(n)} \to Y^{(n+1)}$ is the mapping cone of a map

$$k^{(n)}\colon M(H_{n+1}(C(f),Z),n) \to Y^{(n)}$$

where $M(G,n)$ is the type G-n Moore space:

$$\tilde{H}_m(M(G,n),Z) = \begin{cases} 0 & m \neq n \\ G & m = n \end{cases}$$

and $C(f)$ is the cone on f.

The construction of $Y^{(n)}$ is done by induction: If $f^{(n)}\colon Y^{(n)} \to Y$ is constructed let $V^{(n)} \to Y^{(n)}$ be fiber $f^{(n)}$. Assume inductively that $V^{(n)}$ is $n-1$ connected and $H_n(V^{(n)},Z) = H_{n+1}(C(f),Z)$. Then

$\pi_n(V^{(n)}) = H_{n+1}(C(f),Z)$. Let $M^{(n)} = M(H_{n+1}(C(f),Z),n) \to V^{(n)}$ yield an isomorphism of H_n. $k^{(n)}$ is then $M^{(n)} \to V^{(n)} \to Y^{(n)}$ and $Y^{(n+1)} = C(k^{(n)})$.

If $* = X$ then $f^{(n)}: Y^{(n)} = H\ell_n(Y) \to Y$ is called the *homology (or Moore) approximation of* Y *in* $\dim \leq n$.

While Ht_n is a fairly natural construction and $Ht_n(f)$ is uniquely defined for any map f $H\ell_n$ does not share these conveniences: There is no way one can choose universally maps $H\ell_n(X) \to X$ so that maps $f: X \to Y$ will induce maps $H\ell_n(f): H\ell_n(X) \to H\ell_n(Y)$. e.g. If $|G| < \infty$, $X = Y = M(G,n) \vee S^{n+1}$ $H\ell_n(X) = H\ell_n(Y) = M(G,n)$ for any choices $M(G,n) \to X$, $M(G,n) \to Y$ one can find $f: X \to Y$ so that $f(M(G,n))$ cannot be deformed into $M(G,n)$. Thus one should be very careful in using the $H\ell$ construction.

<u>2.3.1. Proposition (see [Copeland]$_2$ and [Stasheff]$_1$):</u>

Let $f: X,\mu \to X',\mu'$ *be an* H-*map. Suppose* $H_m(f,Z)$ *is an isomorphism for* $m < n$ *and an epimorphism for* $m = n$.

Given any map $g: X' \to K(A,n+1)$ *so that* $gf \sim *$ *then*

(a) g *is an* H-*map.*

(b) *Let* $r: \widetilde{X} \to X'$ *be the fiber of* g. *Then any* H-*structure* $\widetilde{\mu}$ *for* $\widetilde{X}$ *so that* r *is a* $\widetilde{\mu} - \mu'$ H-*map is* (2.1.1) *induced by* μ' *and some* H-*structures of* g. *(The* H-*structure* μ_K *of* $K(A,n+1)$ *is unique as* $[K(A,n+1) \wedge K(A,n+1), K(A,n+1)] = *$.)

(c) *For every* **H**-*structure* F *for* f *and any lifting* $\widetilde{f}: X \to \widetilde{X}$ *of* f *there exists a unique* H-*structure* G *for* g *so that* $\widetilde{F}$ *exists* $\widetilde{f},\widetilde{F}$ *an* **H**-*lifting of* f,F

Proof: (a) $* = HD(*) = HD(gf) = HD(g)(f \wedge f)$ by 1.4.2 (a),

$$HD(g) \in [X' \wedge X', \ K(A,n+1)] = H^{n+1}(X' \wedge X',A).$$

The conditions on $H_m(f,Z)$ imply that $C(f)$ is n connected and consequently $C(f \wedge f)$ is at least $n+1$ connected and hence $H^{n+1}(f \wedge f, A)$ is a monomorphism and $H^{n+1}(f \wedge f, A)HD(g) = 0$ implies $HD(g) = *$.

(b) Choose any H-structure G for g. Let $\tilde{\mu}_0$ be the H-structure induced by μ' and G. Let $\tilde{\mu}$ be any other multiplication in $\tilde{X}$ so that r is a $\tilde{\mu} - \mu'$ H-map. Put

$$w = HD(1,\tilde{\mu},\tilde{\mu}_0): \ \tilde{\mu} = \mu(w \times \mu)\Delta_{\tilde{X} \times \tilde{X}}.$$

As r is both a $\tilde{\mu}_0 - \mu'$ and a $\tilde{\mu} - \mu'$ H-map

$$rw = rHD(1,\tilde{\mu},\tilde{\mu}_0) = HD(r,\tilde{\mu},\mu') = *$$

$$w \in \mathrm{im}(j_*: [\tilde{X} \wedge \tilde{X}, \ \Omega K(A,n+1)] \to [\tilde{X} \wedge \tilde{X}, \ \tilde{X}])$$

where $j: \Omega K(A,n+1) \to \tilde{X}$ is the inclusion of the fiber of r. Now the n-connectivity of $C(r)$ implies that $C(r \wedge r)$ is at least $n+1$ connected and $H^n(r \wedge r, A)$ is an isomorphism. Let

$$w = j_*(r \wedge r)^* v = jv(r \wedge r) \qquad v \in [X' \wedge X', \ \Omega K(A,n+1)]$$

One can easily check that $\tilde{\mu}$ is (2.1.1) induced by

$$P\mu_k(v\Lambda \times G)\Delta_{X' \times X'} \in [X' \times X', \ PK(A,n+1)]$$

(c) Choose any F,G and $\xi: * \sim gf$ (which is equivalent to the choice of $\tilde{f}$), so that f,F g,G are H-maps. By 2.2.1 (b)

$$HD(\tilde{f},\mu,\tilde{\mu}) = jD(\xi,F,G) \in [X \wedge X, \ \Omega K(A,n)]$$

where $\tilde{\mu}$ is induced by g (see part (b)). Now, $H^n(f \wedge f, A)$ is an isomorphism. Let $-w = [H^n(f \wedge f, A)]^{-1}$ $(D(\xi,F,G)) \in [X' \wedge X', K(A,n)]$. Then by 2.2.1 (a) $D(\xi,F,P\mu_k(-w\Lambda \times G)\Delta_{X' \times X'}) = *$ hence the existence of $\tilde{F}$. If G and G' both provide an H-lifting $\tilde{f},\tilde{F}$ then one may assume $G' = P\mu_K(w\Lambda \times G)\Delta_{X' \times X'}$ and by 2.2.1 (a)

$$0 = D(\xi,F,G') = D(\xi,F,G) + w(f \wedge f) = w(f \wedge f)$$

As $H^n(f \wedge f, A)$ is an isomorphism $w \sim *$ and $G \sim G'$ corel E_0, E_∞.

2.3.2. Corollary ([Stasheff]$_1$): *If* $f: X,\mu \to X',\mu'$ *is an H-map then its Postnikov decomposition and primitive Postnikov decompositions consist of H-spaces and H-maps and all the H-structures are uniquely determined by those of* X, X' *and* f.

2.3.3. Remark: Note that the Postnikov decomposition is not uniquely defined as all intermediate spaces can be replaced by their homotopy equivalents, and all maps altered accordingly. 2.3.2 states that once the decomposition is chosen, the H-structures are unique.

2.4. Actions, H-actions and principal fibrations

The notions of actions and principal fibrations used in the category of topological spaces can be adapted to the homotopy category so that some of the homotopy properties of the topological models are preserved. This can be done in various ways. A detailed study of such adaptations can be found, e.g., in [Stasheff]$_2$, [Stasheff]$_5$, [Fuchs] and [Porter]$_2$.

A topological (left) action of a topological group G on a space X is a map $\eta: G \times X \to X$ so that $\eta(g_1 \cdot g_2, x) = \eta(g_1, \eta(g_2,x))$ and $\eta(1,x) = x$.

The weakest homotopy analogue of an action is the following:

2.4.1. Definition: Let X,μ be an H-space and Y a space. *An action of* X,μ *on* Y is a map $\eta: X \times Y \to Y$ so that $\eta(*,y) = y$ and if $h: X \to Y$ is given by $h(x) = \eta(x,*)$ then the following diagram is commutative:

2.4.1.1

$$\begin{array}{ccc} X \times X & \xrightarrow{\mu} & X \\ \downarrow{\scriptstyle 1 \times h} & & \downarrow{\scriptstyle h} \\ X \times Y & \xrightarrow{\eta} & Y \end{array}$$

Note that 2.4.1.1 coincides with 2.1.3 (2) and hence 2.1.4 states that if X,μ acts on Y then the fiber of h admits a multiplication, so that the projection on Y is multiplicative.

If Y,μ_Y is an H-space, an action η of X,μ_X on Y,μ_Y is said to be an *H-action* if in addition

2.4.1.2

$$\begin{array}{ccc} X \times Y \times Y & \xrightarrow{\eta \times 1} & Y \times Y \\ \downarrow{\scriptstyle 1 \times \mu_Y} & & \downarrow{\scriptstyle \mu_Y} \\ X \times Y & \xrightarrow{\eta} & Y \end{array}$$

is commutative.

2.4.2. Lemma: *If* X,μ *H-acts on* Y,μ_Y *then* $h: X \to Y$, $h(x) = \eta(x,*)$ *is an H-map.*

Proof: Let $g: X \times X \to X \times Y \times Y$ be given by $g(x_1,x_2) = x_1,*,h(x_2)$. Then $(1 \times \mu_Y)g = 1 \times h$, $(\eta \times 1)g = h \times h$. By 2.4.1.1 and 2.4.1.2 $\mu_Y(h \times h) = \mu_Y(\eta \times 1)g = \eta(1 \times \mu_Y)g = \eta(1 \times h) = h\mu_X$.

<u>2.4.3. Examples</u>: The classical examples of actions are those equivalent to a group action.

If $f: X \to Y$ is a map, then ΩY acts on the fiber of f in the classical way.

One H-space acts on another via an H-map: $h: X,\mu \longrightarrow X',\mu'$, $\eta(x,x') = \mu'(h(x),x')$. If X',μ' is HA then this is an H-action.

Given any space Y. Suppose Y is $k-1$ connected and $\pi_m(Y) = 0$ for $m > n$.

If ℓ satisfies $k+\ell - 1 > n$, $2\ell > n$ then any $\ell-1$ connected H-space X,μ acts on Y. Furthermore, $h: X \to Y$ defines the action uniquely. (For this statement not to be trivial, one has to assume $\ell \leq n$ otherwise $[X,Y] = *$).

Indeed, as $[X \times Y, Y] \to [X \vee Y, Y]$ is an isomorphism $F(h \vee 1)$ can be extended uniquely to $\eta: X \times Y \to Y$ and as $[X \times X, Y] \to [X \vee X, Y]$ is 1-1 (2.4.1.1) holds. Hence for any map $h: X \to Y$ fiber h is an H-space.

<u>2.4.4. Definition</u>: A map $h: X \to Y$ is a *principal fibration* if h is equivalent to a fiber of a map $g: Y \to B$. I.e.:

$$\begin{array}{ccccc} X & \xrightarrow{\approx} & \tilde{X} & \longrightarrow & LB \\ & {\scriptstyle h}\searrow & \downarrow & & \downarrow \\ & & Y & \xrightarrow{g} & B \end{array}$$

is commutative where $\tilde{X} \to Y$ is the fibration induced by g from E_∞.

Let

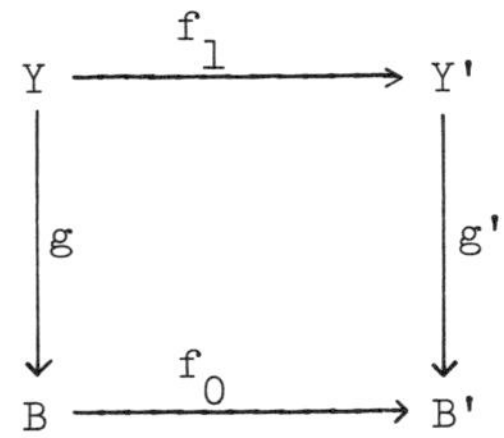

be a commutative diagram and let $F: Y \to PB'$ be a homotopy $f_0 g \sim g' f_1$. Then f_1, f_0 and F induce a map $f = f_{f_0, f_1, F}$: fiber $g \to$ fiber g' $f(y,\gamma) = f_1 y, \; Lf_0\gamma + Fy$.

Let $h: X \to Y$ and $h': X' \to Y'$ be principal fibrations induced by $g: Y \to B$, $g': Y' \to B'$ respectively. *A map of principal fibrations* is a pair $f_1, \hat{f}, f_1: Y \to Y'$, $\hat{f}: X \to X'$ for which maps $f_0: B \to B'$ and $F: Y \to PB'$ $F: f_0 g \sim g' f_1$ exist so that

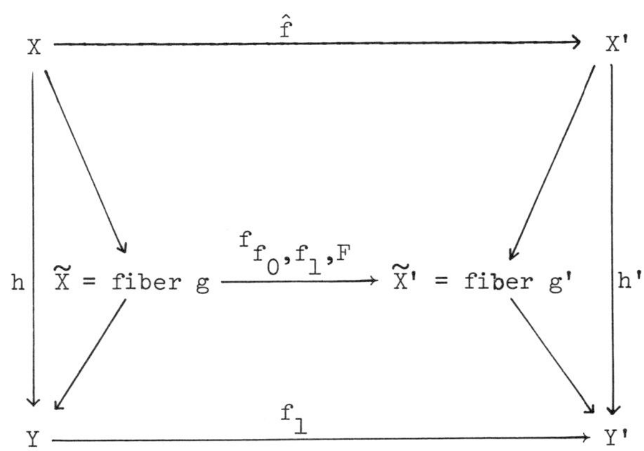

is commutative.

If $f_1, \hat{f}$ is a map of principal fibration $f_1, \hat{f}: (X,Y,h) \to (X',Y',h')$ then the following is commutative

$$\begin{array}{ccc} \Omega B \times X & \xrightarrow{\Omega f_0 \times \hat{f}} & \Omega B' \times X' \\ \downarrow{\scriptstyle \eta} & & \downarrow{\scriptstyle \eta'} \\ X & \xrightarrow{\hat{f}} & X' \end{array}$$

If $h: X \to Y$ is principal induced by $g: Y \to B$ then for every M the group $[M, \Omega B]$ acts on the set $[M,X]$ by an action η_* induced by $\eta: \Omega B \times X \to X$. η_* has the property: For $g_1, g_2 \in [M,X]$ $hg_1 \sim hg_2$ if and only if there exists $\nu \in [M, \Omega B]$ so that $g_2 = \eta_*(\nu, g_1)$.

One can easily see that if $j: L \to M$, $g: M \to X$ $\nu: M \to \Omega B$ then $\eta_*(\nu j, gj) = \eta_*(\nu, g)j$.

If $f_1, \hat{f}: (X,Y,h) \to (X',Y',h')$ is a map of principal fibrations, h,h' induced by $g: Y \to B$ and $g': Y' \to B'$ respectively and $f_1, \hat{f}$ induced by f_0 then for any $g \in [M,X]$, $\nu \in [M, \Omega B]$ one has $\hat{\eta}_*(\Omega f_0 \nu, \hat{f}g) = \hat{f}_* \eta_*(\nu, g)$.

2.4.5. Example:

(a) Every map $h: X \to Y$ with fiber $K(G,n)$ is a principal fibration

(b) Given a commutative diagram

$$\begin{array}{ccc} X & \xrightarrow{f} & X' \\ \downarrow{\scriptstyle h} & & \downarrow{\scriptstyle h'} \\ Y & \xrightarrow{f_1} & Y' \end{array}$$

with h principal induced by $Y \xrightarrow{g} B$, B n-connected, fiber $h' = K(G',n)$ and $H^n(h,G')$ surjective. Then f, f_1 is a map of principal fibrations.

<u>Proof</u>: (a) By the relative Hurewitz theorem $H_m(C(h),Z) = 0$ for $m \leq n$ where $C(h)$ is the cone on h. Let $j: C(h) \to K(H_{n+1}(C(h),Z), n+1)$ be the Postnikov approximation in $\dim \leq n+1$. Let

$$g: Y \to K(H_{n+1}(C(h),Z), n+1)$$

be the composition

$$Y \to C(h) \to K(H_{n+1}(C(h),Z), n+1),$$

$\tilde{X}$ = fiber g and lift h to $\alpha: X \to \tilde{X}$. One may assume that the following commute:

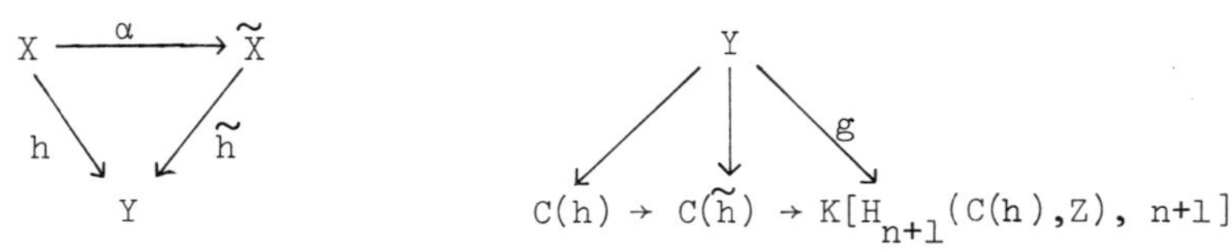

As we assume $\tilde{X}, X$ to be 1-connected by the Serre exact sequence (and see, e.g.: [Spanier] 9.3.4.)

$$H_m(C(\tilde{h}),Z) \to H_m[K(H_{n+1}(C(h),Z), n+1), Z].$$

is an isomorphism for $m \leq n+2$ and as $H_m(j)$ is an isomorphism for $m \leq n+1$ and an epimorphism for $m = n+2$ it follows that $H_m(C(h),Z) \to H_m(C(\tilde{h}),Z)$ is an isomorphism for $m \leq n+1$ and an epimorphism for $m = n+2$. Consequently $H_m(\alpha,Z)$ is an isomorphism for $m \leq n$ and an epimorphism for $m = n+1$, fiber α is n connected and so is the fiber of the map induced by α

$$\hat{\alpha}: \quad \text{fiber } h = K(G,n) \to \text{fiber } \tilde{h} = K(H(C(h),Z), n+1)$$

Hence, $H_{n+1}(C(h),Z) = G$, $\hat{\alpha}$ is a homotopy equivalence and so is α.

(b) Consider the following:

$$\begin{array}{ccc}
X & \xrightarrow{f} & X' \\
\downarrow h & & \downarrow h' \\
Y & \xrightarrow{f_1} & Y' \\
\downarrow g & & \downarrow g' \\
B & & K(G', n+1)
\end{array}$$

By the Serre exact sequence

$$H^{n+1}(B,G') \xrightarrow{g^*} H^{n+1}(Y,G') \xrightarrow{h^*} H^{n+1}(X,G')$$

as $h^*[g'f_1] = H^{n+1}(h,G')[g'f_1] = [g'h'f] = 0$, $[g'f_1] \in \mathrm{im}\, H^{n+1}(g,G')$ and there exists $f_0\colon B \to K(G', n+1)$ so that $g'f_1 \sim f_0 g$. Let $F\colon Y \to PK(G', n+1)$ be such a homotopy. Then $\tilde{f} = f_{f_0,F}$ satisfy $h'\tilde{f} = h'f$, hence, there exists $u\colon X \to K(G',n) = \Omega K(G', n+1)$ with $\eta_*(u,\tilde{f}) = f$. As $H^n(h,G')$ is surjective $u = \nu h$, $\nu\colon Y \to \Omega K(G', n+1)$. Put $\hat{F} = P\mu_K(\nu \times F)\Delta_Y$ then one can easily see that $f_{f_0,\hat{F}} = \eta_*(\nu h, \tilde{f}) = f$.

If $h\colon X \to Y$ is principal induced by $g\colon Y \to B$ then the action $\eta\colon \Omega B \times X \to X$ is equivalent to a group action. As one has a weaker notion of an action, one can analogously seek a weaker form of principal fibrations.

<u>2.4.6. Definition</u>: Given a map $f\colon X \to Y$ with fiber $j\colon F \to X$, f is said to be *weakly principal (abr. w-principal)* if given the pull back diagrams:

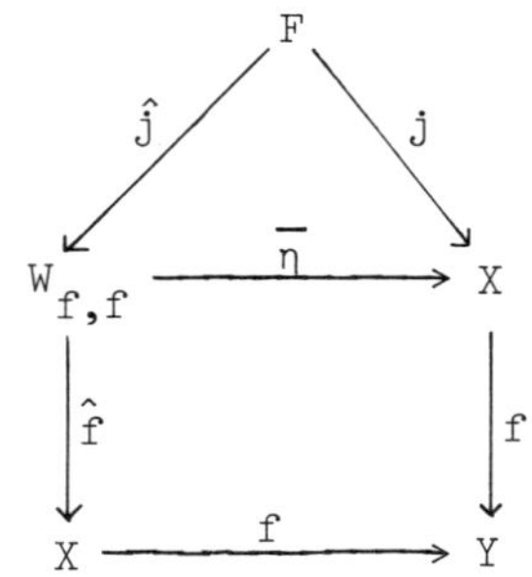

there exists $r\colon W_{f,f} \to F$, $r\hat{j} \sim 1$, $r\chi \sim *$ where $\chi\colon X \to W_{f,f}$ is the obvious cross-section $\chi(x) = x,kx,x \in W_{f,f} \subset X \times PY \times X$.

2.4.7. Lemma: *Given* $f\colon X \to Y$ *with fiber* $j\colon F \to X$. f *is weakly principal if and only if there exists a homotopy equivalence* $u\colon F \times X \to W_{f,f}$ *of type* $\hat{j},\chi$, i.e.: $u(a,*) = \hat{j}(a)$ *and* $u(*,x) = \chi(x)$.

Proof: If $u\colon F \times X \xrightarrow{\approx} W_{f,f}$ of type $\hat{j},\chi$ exists, let $v\colon W_{f,f} \to F \times X$ be its inverse then $r = p_1 v\colon W_{f,f} \to F$ satisfies $r\hat{j} \sim 1$, $r\chi \sim *$.

Suppose conversely that $r\colon W_{f,f} \to F$ exists, $r\hat{j} \sim 1$, $\chi\hat{j} \sim *$. Then one can easily see that

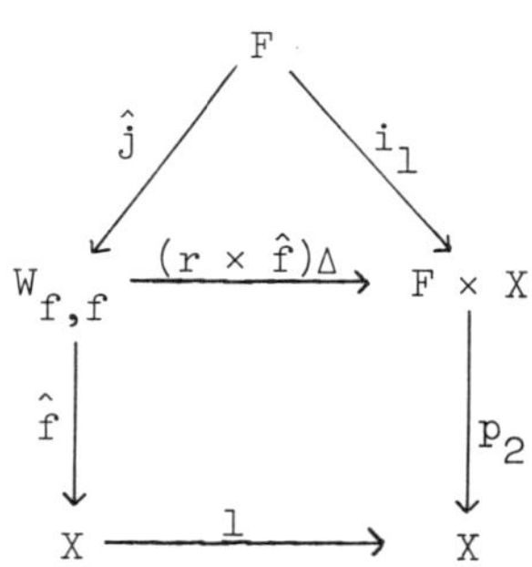

is a fiber homotopy equivalence. As the composition

$F \vee X \xrightarrow{\hat{j} \vee \chi} W_{f,f} \xrightarrow{(r \times \hat{f})\Delta} F \times X$ is homotopic to the inclusion, and $(r \times \hat{f})\Delta$ is a homotopy equivalence, $\hat{j} \vee X$ extends to $u\colon F \times X \to W_{f,f}$ of type $\hat{j}, X$ homotopic to the inverse of $(r \times \hat{f})\Delta$.

<u>2.4.8. Corollary</u>: *If* $f\colon X \to Y$ *is w-principal, then one has a pull back diagram*

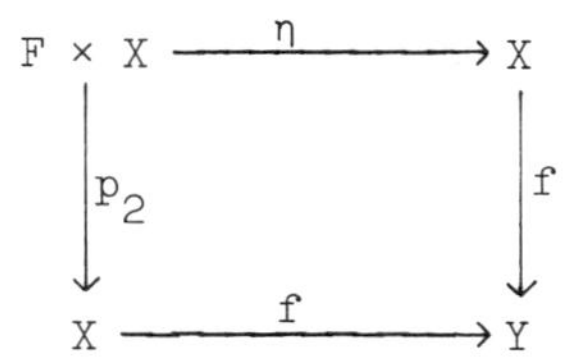

with η *of type* $j,1$.

<u>Proof</u>: In the notations of 2.4.7 and its proof, put $\eta = \bar{\eta}u$ and note that $\hat{f}u = p_2 vu \sim p_2$ and $f\eta = f\bar{\eta}u = f\hat{f}u = fp_2$.

<u>2.4.9. Proposition</u>: *Let* $f\colon X \to Y$ *be w-principal. Given* $g\colon Y' \to Y$ *and let* $f'\colon X' = W_{g,f} \to Y'$ *be the fibration induced from* f *by* **g**. *Then* $f'\colon X' \to Y'$ *is w-principal.*

<u>Proof</u>: One can see that if $\tilde{g}\colon X' \to X$ covers g then one has a pull back diagram

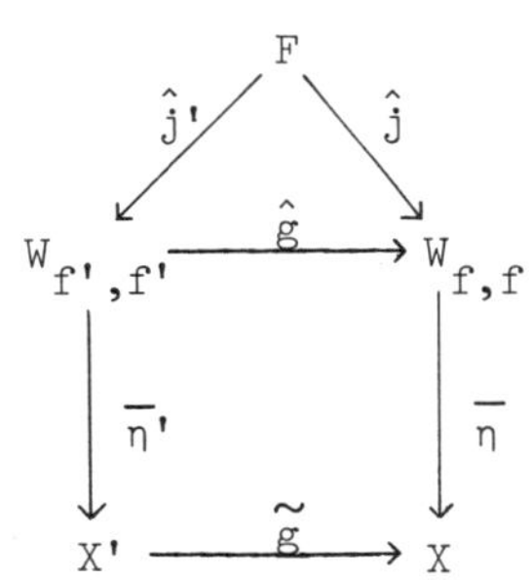

(This can be easily seen if one replaces $f\colon X \to Y$ by a fibration and the homotopy pull backs by topological ones.)

Furthermore, if $\chi_{X'}\colon X' \to W_{f',f'}$ then $\chi_X \tilde{g} = \hat{g}\chi_{X'}$. If $r\colon W_{f,f} \to F$ satisfies $r\hat{j} \sim 1$, $r\chi_X \sim *$ then $r\hat{g}\colon W_{f',f'} \to F$ satisfies the corresponding properties.

A justification for the definition of w-principality is given by

<u>2.4.10. Proposition</u>: *If* $f\colon X \to Y$ *is w-principal, then* $F = \text{fiber } f$ *is an H-space which* η*-acts on* X. *This action induces an action* $\eta_*\colon [M,F] \times [M,X] \to [M,X]$ *so that for* $g_1, g_2\colon M \to X$ $fg_1 \sim fg_2$ *if and only if* $g_2 = \eta_*(u,g_1)$ *for some* $u \in [M,F]$.

(Note that $[M,F]$ *is not necessarily a group and its action on the set* $[M,X]$ *does not satisfy any associativity properties. The only properties hold are* $1\cdot x = x$ *and* $(u_1\cdot u_2)\cdot x_0 = u_1(u_2\cdot x_0)$ *for the base point* $x_0 = * \in [M,X]$.)

<u>Proof</u>: Consider the following commutative diagram:

$$\begin{array}{ccc} F \times F & \xrightarrow{\ \bar{m}\ } & F \\ \downarrow{\scriptstyle 1\times j} & & \downarrow{\scriptstyle j} \\ F \times X & \xrightarrow{\ \eta\ } & X \\ \downarrow{\scriptstyle p_2} & & \downarrow{\scriptstyle f} \\ X & \xrightarrow{\ f\ } & Y \end{array}$$

As $f\eta(1 \times j) \sim fp_2(1 \times j) = fjp_2 \sim *$ $\eta(1 \times j)$ lifts to $\bar{m}\colon F \times F \to F$. $j\bar{m}|F \vee F \sim \eta(1 \times j)|F \vee F = (\eta|F \vee X)(1 \vee j) \sim j \vee j = jF$. As $\Omega Y \to F \xrightarrow{j} X$ is principal with action say $\hat{\eta}$, there exists $\alpha\colon F \vee F \longrightarrow \Omega Y$ so that $F = \hat{\eta}_*(\alpha, \bar{m}|F \vee F)$. α extends to $\bar{\alpha}\colon F \times F \longrightarrow \Omega Y$ and if one replaces $\bar{m}$ by $\mu = \hat{\eta}_*(\bar{\alpha},\bar{m})$ then $\mu|F \vee F = \hat{\eta}_*(\bar{\alpha},\bar{m})|F \vee F = \hat{\eta}_*(\bar{\alpha}|F \vee F, \bar{m}|F \vee F) = \hat{\eta}_*(\alpha,\bar{m}|F \vee F) = F$.

Obviously $j\mu = j\overline{m} = \eta(1 \times j)$ and η is an action.

2.4.11. Corollary: *Let* $f\colon X \to Y$ *be a map of spaces.* f *is weakly principal if and only if* $F = \text{fiber } f$ *is an H-space* η*-acting on* X *with* $\eta(a,*) = j(a)$, $j\colon F \to X$ *and in such a way that the obvious homotopy* $f\eta|F \vee X \sim fp_2|F \vee X$ *can be extended to a homotopy* $H\colon f\eta \sim fp_2$.
(*Note that by* 1.1.7 *if* Y *is an H-space, it suffices to assume that* $F = \text{fiber } f$ *is an H-space* η*-acting on* X *with* $f\eta \sim fp_2$).

Proof: If f is w-principal, then $\eta = \overline{\eta}u$ (with the notation of 2.4.6 and 2.4.7) satisfies the hypothesis of this proposition as can be seen directly from the definitions of the various spaces and maps.

If $\eta\colon F \times X \to X$ with the desired properties exists, let $u\colon F \times X \to W_{f,f}$ be given by $u(a,x) = \eta(a,x), H(a,x),x$, as $\hat{f}u = p_2$ $u(a,*) = j(a), H(a,*),*$ and if $a = x,\gamma \in X \times LY$ $j(a) = x$, by our hypothesis, $H(a,*) = \gamma$, hence $u(a,*) = \hat{j}(a)$ and u is a homotopy equivalence. As $u(*,x) = x$, $H(*,x),x$ and again by the hypothesis $H(*,x) = kf(x)$ $u(*,x) = \chi(x)$ and u is of type $\hat{j},\chi$ and 2.4.11 follows from 2.4.7.

2.4.12. Example: Let X,μ right $\hat{\eta}$-act on Y with $h(x) = \hat{\eta}(*,x)$. Then h satisfies the hypothesis of 2.4.11 and hence is w-pricipal. Indeed, $F = \text{fiber } h$ admits an H-structure $\hat{\mu}$ with $j\colon F \to X$ multiplicative. Let $\eta\colon F \times X \to X$ be given by $\mu(j \times 1)$. Then one has a commutative diagram

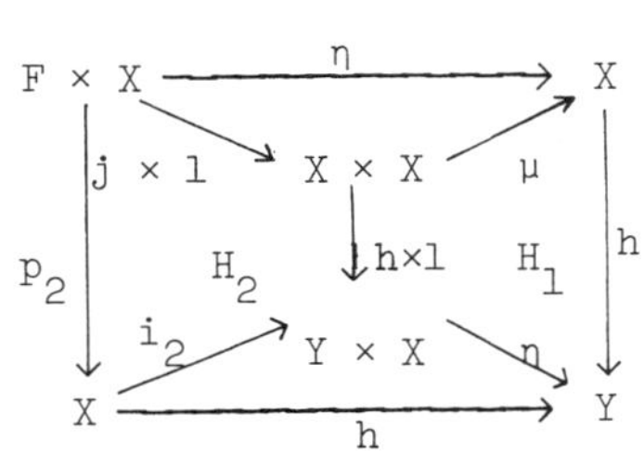

The homotopy $h\eta \sim hp_2$ can be given by $H = H_1(j \times 1) + P\hat{\eta}H_2$ where H_1 is a homotopy $h\mu \sim \hat{\eta}(j \times 1)$ rel $X \vee X$ and H_2: $hj \times 1 \sim * \times 1 = i_2p_2, H_2 = \hat{H}_2 \times 1$, $\hat{H}_2$: $F \to PY$, the obvious homotopy $hj \sim *$. One can easily see that $H|F \vee X$ is the obvious homotopy.

2.5. HA and HC obstructions

As in the case of the study of H-maps where the homotopy was incorporated as a structural part of the map to obtain the category H_{fF}, it is sometimes convenient to carry the homotopy $\mu(\mu \times 1) \sim \mu(1 \times \mu)$ as part of the structure of an HA space. Thus:

2.5.1. Definition: *An HA space (in the precise sense)* is a triple X,μ,A where X,μ is an H-space and A: $X \times X \times X \longrightarrow PX$ is a homotopy $\mu(\mu \times 1) \sim \mu(1 \times \mu)$ which by 1.1.5 and 1.1.6 can be assumed to be rel $X \vee X \vee X = \overset{2}{\underset{3}{\vee}} X$.

An A-map (A_3 map in the sense of [Stasheff]$_2$) is a triple f,F,a: $X,\mu,A \to X',\mu',A'$ so that (f,F) is an H-map $X,\mu \to X',\mu'$ and a: $X \times X \times X \longrightarrow P(PX')$ satisfies

$$(PE_0)a = F(\mu \times 1) + P\mu'(F \times kf), \quad E_0 a = (Pf)A$$

$$(PE_\infty)a = F(1 \times \mu) + P\mu'(kf \times F), \quad E_\infty a = A'(f \times f \times f).$$

a can be illustrated as follows:

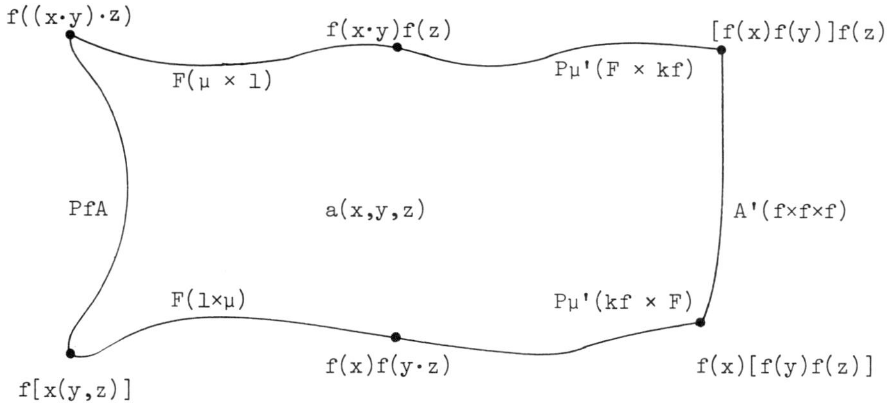

Let $f,F\colon X,\mu \to X',\mu'$ be an H-map, X,μ,A and X',μ',A' HA spaces. Let $\tilde{\theta} = \tilde{\theta}(f,F,A,A')$, $\tilde{\theta}\colon X \times X \times X \longrightarrow (X')^{S^1} \subset PX'$ be given by

$$\tilde{\theta} = P(\mu\times 1) + P\mu'(F \times kf) + A'(f\times f\times f) - P\mu'(kf \times F) - F(1\times\mu) - PfA.$$

$\tilde{\theta}|X \veebar X \veebar X \sim kf(\mu \times 1)|X \veebar X \veebar X$ corel E in an obvious way,
$\tilde{\theta} \sim \tilde{\theta}'$ corel E with $\tilde{\theta}'|X \veebar X \veebar X = kf(\mu \times 1)|X \veebar X \veebar X$.

Let $\theta = \theta(f,F,A,A')\colon X \wedge X \wedge X \to \Omega X'$ be given by $\theta\Lambda^3 = D_{\tilde{\theta}',kf(\mu \times 1)}$ then θ is the obstruction for f,F to be an A-map. The way θ is affected by composition is given by

<u>2.5.2. Proposition</u>: *Let* $(f,F)\colon X,\mu \to X',\mu'$ $(f',F')\colon X',\mu' \to X'',\mu''$ *be H-maps where* X,μ,A, X',μ',A' *and* X'',μ'',A'' *are HA spaces. Then* $\theta((f',F')(f,F),A,A'') = \Omega f'\theta(f,F,A,A') + \theta(f',F',A',A'')(f \wedge f \wedge f)$.

<u>Proof</u>: The proof is illustrated by

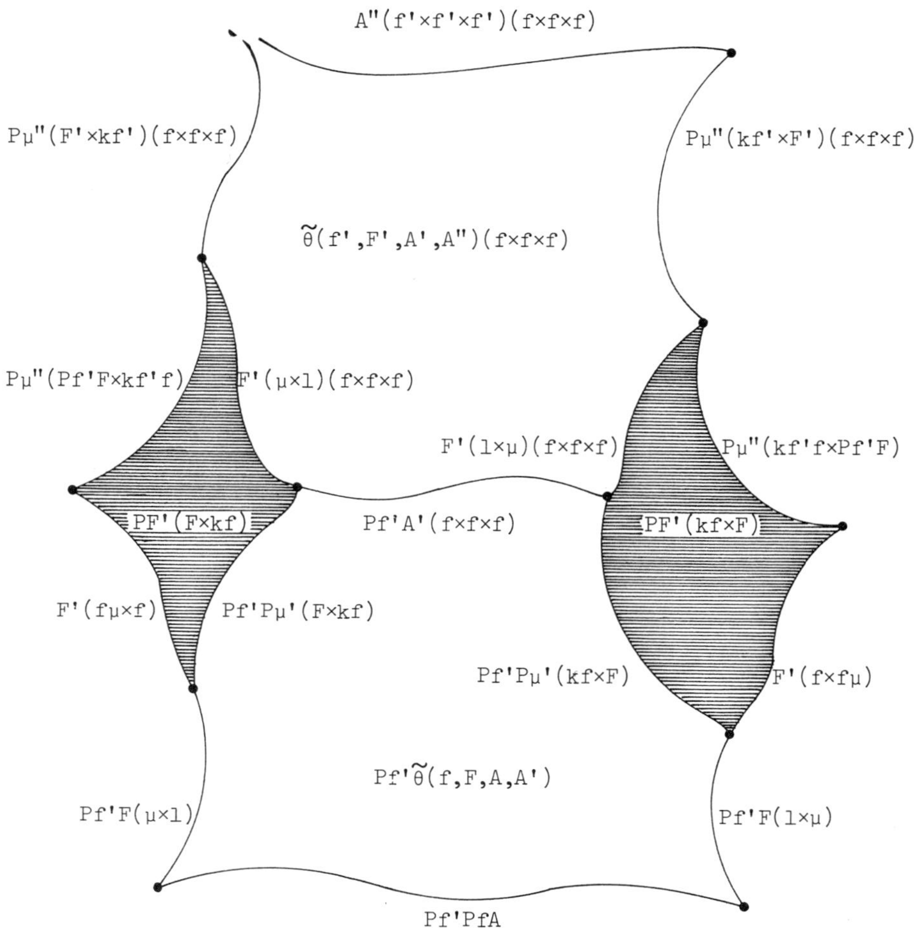

<u>2.5.3. Proposition</u>: *Let* $f_i, F_i, a_i\colon\ X_i, \mu_i, A_i \longrightarrow X_0, \mu_0, A_0$ $i = 1,2$ *be HA maps. Then* a_i, A_i *induce an HA structure on the pull back* W_{f_1,f_2} *of* f_1 *and* f_2 *with* 2.1.1 *H-structure.*

<u>Proof</u>: Let $z = (x_1, \bar{\gamma}, x_2)$, $z' = (x_1', \bar{\gamma}', x_2')$ and $z'' = (x_1'', \bar{\gamma}'', x_2'')$ be points in W_{f_1,f_2}. Then

$$\hat{\mu}(z,z') = \mu_1(x_1,x_1'),\ F_1(x_1,x_1') + P\mu_0(\gamma,\overline{\gamma}) - F_2(x_2,x_2'),\ \mu_2(x_2,x_2') =$$

$$= x_1\cdot x_1',\ F_1(x_1,x_1') + \gamma\cdot\overline{\gamma} - F_2(x_2,x_2'),\ \ x_2\cdot x_2'$$

$$\hat{\mu}(\hat{\mu}(z,z'),z'') = (x_1\cdot x_1')x_1'',\ F_1(x_1\cdot x_1',\ x_1'') +$$

$$+ [F_1(x_1,x_1') + \overline{\gamma}\cdot\overline{\gamma}' - F_2(x_2,x_2')]\cdot\overline{\gamma}'' -$$

$$- F_2(x_2\cdot x_2',\ x_2''),\ (x_2\cdot x_2')x_2''$$

and using 1.2.4.

$$(z\cdot z')z'' = (x_1\cdot x_1')x_1'',\ F_1(x_1\cdot x_1',x_1'') + F_1(x_1,x_1')kf_1(x_1'') + (\overline{\gamma}\cdot\overline{\gamma}')\overline{\gamma}'' -$$

$$- F_2(x_2,x_2')kf_2(x_2'') - F_2(x_2\cdot x_2',x_2''),\ (x_2\cdot x_2')x_2''$$

Similarly,

$$z(z'\cdot z'') = x_1(x_1'\cdot x_1''),\ F_1(x_1,x_1'\cdot x_1'') +$$

$$+ kf_1(x_1)\cdot F_1(x_1',x_1'') + \overline{\gamma}(\overline{\gamma}'\cdot\overline{\gamma}'') -$$

$$- kf_2(x_2)\cdot F_2(x_2',x_2'') - F_2(x_2,\ x_2'\cdot x_2''),\ x_2(x_2'\cdot x_2'').$$

The homotopy $(z\cdot z')z'' \sim z(z'\cdot z'')$ is illustrated by

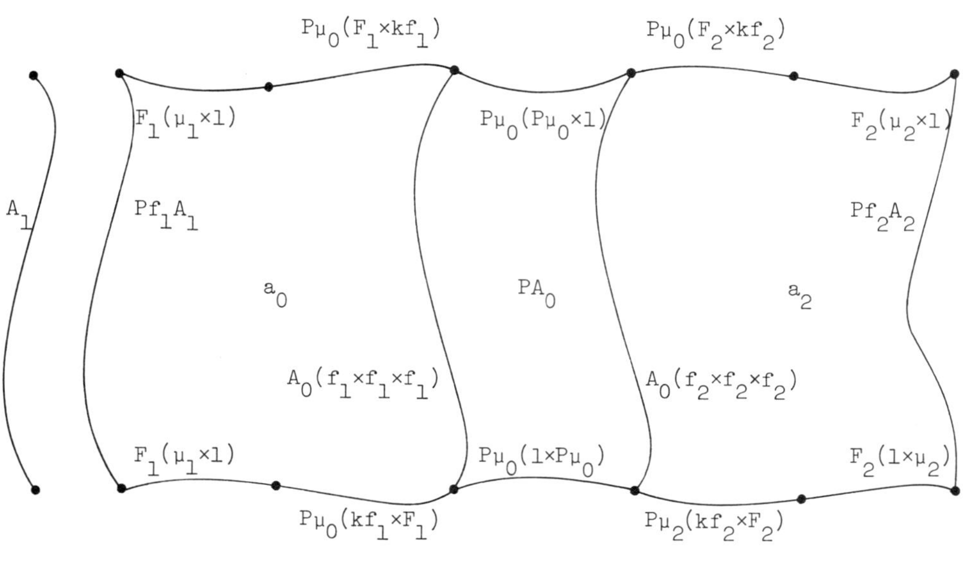

One can carry out the same procedure for HC spaces. The proofs involve two rather than three variables, are simpler and hence omitted. (See [Araki, Kudo] and $[\text{Browder}]_2$.)

An HC space can be regarded as a triple X,μ,C where X,μ is an H-space and $C\colon X \times X \longrightarrow PX$ is a homotopy $\mu \sim \mu T$ and one may assume that $C|X \vee X = kF$.

An HC map is a triple $f,F,c\colon\ X,\mu,C \to X',\mu',C'$ so that f,F is an H-map and $c\colon X \times X \longrightarrow P(PX')$ is illustrated by

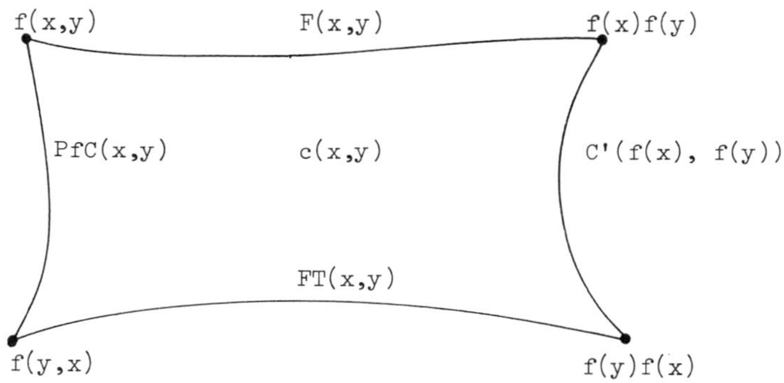

$$(PE_0)c = F \qquad E_0 c = Pfc$$

$$(PE_\infty)c = FT \qquad E_\infty c = C'(f \times f)$$

Let X,μ,C and X',μ',C' be HC spaces, let $f,F\colon\ X,\mu \to X',\mu'$ be an H-map. Define $\tilde{\phi} = \tilde{\phi}(f,F,C,C')\colon\ X \times X \to X'^{S^1} \subset PX'$ by $\tilde{\phi} = F + C'(f \times f) - FT - PfC$ then $D_{\tilde{\phi},kf}$ defines a map $\phi\colon X \wedge X \to \Omega X'$ which is the obstruction for f,F to be an HC map.

<u>2.5.4. Proposition</u>: *If* $f,F\colon\ X,\mu \to X',\mu'$ $\quad f',F'\colon\ X',\mu' \to X'',\mu''$ *are H-maps* $X,\mu,C,$ X',μ',C' *and* X'',μ'',C'' *HC spaces then*

$$\phi((f',F')(f,F),C,C'') = \Omega f'\phi(f,F,C,C') + \phi(f',F',C',C'')(f \wedge f)$$

2.5.5. Proposition: *Let* $f_i, F_i, c_i: X_i, \mu_i, C_i \to X_0, \mu_0, C_0 \quad i = 1,2$ *be HC maps. Then* c_i, C_i *induce an HC structure on the pull back* W_{f_1,f_2} *with the* 2.1.1 *H-structure.*

2.6. Homotopy solvability and homotopy nilpotency.

Given an H-space X,μ one can look for the algebraic properties of $[M,X]$ and in particular those properties which are independent of M and are thus invariants of X,μ. In particular we are interested in the properties of solvability and nilpotency.

Recall that for a group (or an algebraic loop) G solvability of class $< n$ is equivalent to the property that the function (on sets) $\alpha_n: \underbrace{G \times G \times \ldots \times G}_{2^n} \longrightarrow G$ given by $\alpha_2(g_1,g_2) = [g_1,g_2]$,

$$\alpha_n(g_1, g_2, \ldots, g_{2^n}) = [\alpha_{n-1}(g_1,\ldots,g_{2^{n-1}}),\ \alpha_{n-1}(g_{2^{n-1}+1},\ldots,g_{2^n})]$$

is constantly 1. $([g_1,g_2] = (g_1\cdot g_2)(g_2\cdot g_1)^{-1})$.

Similarly G is nilpotent of class $< n$ if the function $\beta_n: G^n \to G$ given by $\beta_n(g_1,\ldots,g_n) = [\ldots[g_1,g_2], g_3],\ldots], g_n]$ is constantly 1.

The question one may ask for a given H-space X,μ is: When is $[M,X]$ solvable (or nilpotent) of class $< n$ for all M? This problem is studied in [Harrison -Scheerer], [Whitehead], [Berstein-Ganea], $[\text{Arkowitz-Curjel}]_{2,3,4}$, and $[\text{Porter}]_1$. Such a property can be transformed into an intrinsic property of X as follows:

2.6.1. Lemma: $[\ ,X]$ *is solvable (nilpotent) of class* $< n$ *if and only if for* $G = [X^{2^n},X]$ $(G = [X^n,X])$ $\alpha_n(p_1, p_2,\ldots,p_{2^n}) = *$ $(\beta_n(p_1,\ldots,p_n) = *)$ *where* p_i *are the projections.*

<u>Proof</u>: It is obvious that $< n$ solvability (nilpotency) of $[M,X]$ for all M implies $\alpha(p_1,\ldots,p_{2^n}) = *$ $(\beta(p_1,\ldots,p_n) = *)$. On the other hand, for any M and $f_i: M \to X$ there exists $\hat{f}: M \to X^{2^n}$ $(\hat{f}: M \to X^n)$ so that $f_i = p_i\hat{f}$. As $\hat{f}^*: [X^{2^n},X] \to [M,X]$ $(\hat{f}^*: [X^n,X] \to [M,X])$ are homomorphisms

$$\hat{f}^*\alpha(p_1,\ldots,p_{2^n}) = \alpha(f_1,\ldots,f_{2^n}) \; (\hat{f}^*\beta(p_1,\ldots,p_n) = \beta(f_1,\ldots,f_n))$$

We are now to define formally $\alpha(p_1,\ldots,p_{2^n})$ and $\beta(p_1,\ldots,p_n)$ and study their fundamental properties.

For an H-space X,μ let $w_2 = w_2(X,\mu) \in [X \wedge X, X]$ $v_1 = v_1(X,\mu) \in [X \wedge X, X]$ be given by $w_2\Lambda = D_{\mu,\mu T} = v_1\Lambda$.

Let $\Lambda^n X$ be the n-fold $\wedge$-product of X, $\Lambda^1 X = X$, $\Lambda^n X = X \wedge \Lambda^{n-1} X$ and define inductively $w_n = w_n(X,\mu)$: $\Lambda^n X \to X$, $w_n = w_2(w_{n-1} \wedge 1)$ $v_n = v_n(X,\mu)$: $\Lambda^{2^n} X \to X$, $v_n = v_1(v_{n-1} \wedge v_{n-1})$.

<u>2.6.2. Definition</u>: X,μ is said to be *homotopy nilpotent of class* $< n$ *(abr:* $< n$ *HN)* if $w_n(X,\mu) \sim *$.

X,μ is said to be *homotopy solvable of class* $< n$ *(abr:* $< n$ *HS)* if $v_n \sim *$.

The following simple properties of w_n and v_n can be easily proved by induction:

<u>2.6.3. Lemma</u>:

(a) $v_{n+m} = v_n(\Lambda^{2^n} v_m)$

(b) If $f: X,\mu \to X',\mu'$ *is an H-map then*

$$fw_n(X,\mu) = w_n(X',\mu')(\Lambda^n f)$$

$$fv_n(X,\mu) = v_n(X',\mu')(\Lambda^{2^n} f)$$

(c) $w_2(X,\mu) = 0$ *if and only if* X,μ *is HC.*

(d) $v_{n-1} = w_n(1 \wedge 1 \wedge v_1 \wedge v_2 \wedge v_3 \wedge \ldots \wedge v_{n-2})$ *and hence* $< n$ HN *implies* $< n-1$ HS.

2.6.3 suggests the following generalizations of 2.6.2:

<u>2.6.4. Definition</u>: Let $f\colon X,\mu \to X'\mu'$ be an H-map. f is said to be $< n$ HN ($< n$ HS) if $fw_n(X,\mu) = *$ ($fv_n(X,\mu) = *$).

<u>2.6.5. Definition</u>: An H-map $f\colon X,\mu \to X',\mu'$ is said to be *central* if $w_2(X',\mu')(f \wedge 1) = *$.

The elementary techniques for study of nilpotency and solvability of groups can apply to the study of HN and HS properties:

<u>2.6.6. Lemma: (see [Berstein, Ganea] and [Larmore, Thomas])</u>

Let $F,\mu_F \xrightarrow{j} E,\mu_E \xrightarrow{f} B,\mu_B$ *be an H-fibration, i.e.:* $F \to E \to B$ *is a fibration and all spaces and maps are H-spaces and H-maps.*

(a) If f *is* $< n$ HN *and* j *is central then* E *is* $< n+1$ HN.

(b) If j *is* $< m$ HS, $f < k$ HS *then* E *is* $< m+k$ HS.

(c) Let $g\colon B,\mu_B \to B_0, \mu_{B_0}$ *be an H-map, considering the induced principal H-fibration* $\Omega B_0, \mu_\Omega \xrightarrow{j} E, \mu_E \xrightarrow{f} B,\mu_B$, μ_E 2.1.1 *induced,* $\mu_\Omega = \text{Add}$. *Then* j *is central.*

<u>Proof</u>: (a) $fw_n(E,\mu_E) = *$ implies $w_n(E,\mu) = j\tilde{w}$ for some $\tilde{w} \in [\Lambda^n E, F]$. $w_{n+1}(E,\mu_E) = w_2(E,\mu_E)[w_n(E,\mu_E) \wedge 1] = w_2(E,\mu_E)(j\tilde{w} \wedge 1) =$ $= w_2(E,\mu_E)(j \wedge 1)(\tilde{w} \wedge 1) = *$ as j is central.

(b) $v_{m+k}(E,\mu_E) = v_m(E,\mu_E)\Lambda^{2^m}(v_k(E,\mu_E))$. $f < k$ HS implies $fv_k(E,\mu_E) \sim *$ and $v_k(E,\mu_E) = j\tilde{v}$, $\tilde{v} \in [\Lambda^{2^k} E, F]$ and as $jv_m(F,\mu_F) = *$

one has $v_{m+k}(E,\mu_E) = v_m(E,\mu_E)\Lambda^{2^m}(v_k(E,\mu_E)) = v_m(E,\mu_E)\Lambda^{2^m}(j)\Lambda^{2^m}\tilde{v} = jv_m(F,\mu_F)\Lambda^{2^m}\tilde{v} = *$.

(c) $E = \{b, \gamma \in B \times LB_0 | g(b) = E_\infty\gamma\}$, $\mu_E[(b,\gamma), (\bar{b},\bar{\gamma})] = \mu_B(b,\bar{b}), L\mu_{B_0}(\gamma,\bar{\gamma}) + G(b,\bar{b})$ where G is an H-structure for g. $j(\lambda) = *,\lambda$

$$\mu_E(j \times 1)(\lambda, (b,\gamma)) = b,\ L\mu_{B_0}(\lambda,\gamma) + kg(b) = b,\ L\mu_{B_0}(\lambda,\gamma)$$

$$\mu_E T(j \times 1)(\lambda, (b,\gamma)) = b,\ L\mu_{B_0}(\gamma,\lambda)$$

By 1.2.4 and 1.2.5 $L\mu_{B_0}T \sim L\mu_{B_0}$ on $\Omega B_0 \times LB_0$ (corel E_0, E_∞) hence j is central.

2.6.7. Corollary (see [Berstein, Ganea]):

(a) *If* X *is an H-space,* $\pi_m(X) = 0$ $m \neq n_1, n_2, \ldots, n_k$, *then* X *is* $< k+1$ HN.

(b) *Let* K *be a connected finite complex* $\dim K = k$. *Then for any H-space* X, X^K, μ^K *is* $< k+1$ HN.

Proof: Both X (in (a)) and X^K (in case (b)) can be obtained via a sequence of k principal H-fibrations: (a) by its Postnikov system and (b) by the principal cofibrations inducing the cellular structure of K (with the bottom space $X^{\tilde{S}}$, $\tilde{S} = \bigvee_m S^1$. $X^{\tilde{S}} = (\Omega X)^m$ is HC).

2.6.8. Proposition (Ganea): ΩS^{2n+1} *is* < 3 HN.

Proof: By [James]$_1$ and its consequences, one has the following James fibration:

$$S^{2n+1} \xrightarrow{j} \Omega S^{2n+2} \xrightarrow{f} \Omega S^{4n+3}$$

Hence, one has a fibration

$\Omega^3 S^{4n+3} \xrightarrow{h} \Omega S^{2n+1} \xrightarrow{j} \Omega^2 S^{2n+2}$ and as $\Omega^2 S^{2n+2}$ is HC

(hence < 2 HN) 2.6.8 follows from 2.6.6 (c) and (a).

2.6.9. Proposition: *Let* $G_n = SU(n)$ *or* $Sp(n)$. *Put* $V_n^m(G) = G_n/G_m$. *Then* $\Omega V_n^m(G)$ *is* $< 2(n-m)$ HS.

Proof: One has a fibration

$$V_k^m(G) \to V_n^m(G) \to V_n^k(G) \qquad m \leq k \leq n$$

and by looping one obtains an H-fibration

$$\Omega V_k^m(G) \to \Omega V_n^m(G) \to \Omega V_n^k(G)$$

As $V_n^{n-1}(G) = S^{dn-1}$ ($d = 2$ if $G = SU$ and $d = 4$ if $G = Sp$) and by 2.6.8 ΩS^{dn-1} is < 3 HN, hence < 2 HS, 2.6.9 follows by induction from 2.6.6 (b).

Now, it is well known that for $G = SO$, SU or Sp the inclusion $G(n) \to G(2n)$ is < 2 HN (or < 1 HS). (This property is stated as $G(n)$ homotopy commutes in $G(2n)$.) As the fiber of $G(n) \to G(2n)$ is $\Omega V_{2n}^n(G)$ and $\Omega V_{2n}^n(G) \to G(n) \to G(2n)$ is an H-fibration with respect to the loops addition and Lie group multiplications, 2.6.9 and 2.6.6 (b) imply:

2.6.10. Proposition: $SU(n)$ ***and*** $Sp(n)$ *are* $< 2n+1$ HS. *Hence, for any topological space* Y *(not necessarily of the homotopy type of a CW complex)* $[Y, SU(n)]$ *and* $[Y, Sp(n)]$ *are* $< 2n+1$ *solvable groups.*

2.6.11. Remark: 2.6.10 can be extended to $SO(2n+1)$ via a primitive treatment as studied in Chapter 4: The James fibration holds mod 2 for S^{2n} as well and for odd primes $SO(2n+1)/SO(2n-1) = S^{4n-1}$.

The argument can be also extended to some exceptional groups, e.g:

Example (Adams): Using the fibration

$$G_2 \to Spin(7) \to S^7$$

one has an H-fibration: $\Omega S^7 \to G_2 \to Spin(7)$. As Spin (7) is < 15 HS (as SO(7)) and ΩS^7 is HC G_2 is < 16 HS.

Note that these methods do not yield the minimum k for which these groups are $< k$ HS and the numbers given here almost certainly can be improved.

Chapter III

The Cohomology of H-spaces

Introduction

The study of the cohomology of H-spaces has, amongst others, the following applications: The fact that a space X admits an H-structure implies the existence of certain properties in its cohomology. This may help to determine the cohomology of a space, knowing that it admits a multiplication. Conversely, if a cohomology of a space is known, the general theory of the cohomology of H-spaces may help to determine whether or not the space admits a multiplication.

In this chapter we mainly study the ordinary cohomology with $Z_p = Z/pZ$ coefficients, though the results of section 2 are given for an arbitrary theory. The same is true for the part underlying the Bockstein spectral sequence in section 3.

If X,μ is an H-space and F is a field then $H^*(X,F)$ together with $H^*(\mu,F)$ and $H^*(\Delta,F)$ form a Hopf algebra. We assume the reader is familiar with the basic notions regarding Hopf algebras and, as usual, the standard reference is [Milnor and Moore] whose notations we adopt freely.

Section 1 is devoted to the consequences derived through the Hopf algebra theory only. In section 2 we study some properties of the coproduct in $H^*(\Omega X,F)$ derived from properties of the algebra $H^*(X,F)$ (Proposition 3.2.3). As was stated above, this is done with no added stress for generalized theories. It is being applied (3.2.4) to determine the Hopf algebra $H^*(Spin(n), Z_2)$.

Section 3 is devoted to the Bockstein spectral sequence, and hence we assume the basic acquaintance with spectral sequences. As stated before, the spectral sequence is constructed for generalized theories, though applications are all classical. Further results are presented here without proof as the complete pursuit of the proofs requires too many technical details.

In section 4 high order cohomology operations are added to the Hopf algebra structure. The main type is that of non-stable secondary operations. In some sense this technique has its origin in $[\text{Thomas}]_{1,2,3}$ though there the secondary operations are replaced by primary operations in the cohomology of the projective space of the given H-space (a subject not covered in these notes). There are some limitations to this method: Not all cohomology classes of the given space appear in the cohomology of its projective space; not all secondary operations in the cohomology of the space have analogues in its projective space; and, finally, this method is effective mostly for the treatment of the cohomology with coefficients in Z_2 while for odd primes it is weaker.

The study of the secondary operations techniques requires some knowledge of the structure of the Steenrod algebra and of the cohomology of Eilenberg MacLane Spaces. For the details regarding the Steenrod algebra, the standard reference is [Steenrod and Epstein].

3.1. The Hopf algebra $H^*(X, Z_p)$:

If X,μ is an H-space, the obvious commutative diagram

$$\begin{array}{ccc} X \times X & \xrightarrow{\Delta_{X \times X}} & X \times X \times X \times X \\ \downarrow \mu & & \downarrow \mu \times \mu \\ X & \xrightarrow{\Delta_X} & X \times X \end{array}$$

implies that for a field F $H^*(X,F)$ together with $H^*(\Delta,F)$ and $H^*(\mu,F)$ form a Hopf algebra. Thus, one can apply the fundamental structure theorem of Hopf algebras for $H^*(X, Z_p)$:

3.1.1. Theorem (Hopf-Borel): *Let* A *be a graded connected Hopf algebra of finite type over a field* F. *If the characteristic of* F *is not* 0 *assume that* F *is perfect. Then as an algebra*

$$A = [\bigotimes_i \Lambda(x_i)] \otimes [\bigotimes_j \Gamma_{r_j}(y_j)], \qquad 0 < r_j \leq \infty.$$

where $\Lambda(x_i)$ *is an exterior algebra on one generator* x_i *and* $\Gamma_{r_j}(y_j) = F[y_j]/(y_j^{p^{r_j}})$, p = charac. F. (*Put* $\Gamma_{r_j}(y_j) = F[y_j]$ *if* charac. $F = 0$ *or* $r_j = \infty$.)

As the multiplication in a Hopf algebra is always assumed to be commutative and associative, if $p \neq 2$: $\dim x_i$ = odd, $\dim y_i$ = even.

Recall that the augmentation ideal of a Hopf algebra A is the ideal $\bar{A} = \ker \varepsilon$, $\varepsilon: A \to F$ the counit. If A is connected, then $\bar{A} = \bigoplus_{i>0} A_i$.

The module QA of indecomposables of an algebra A is the cokernel of $\bar{\phi}: \bar{A} \otimes \bar{A} \longrightarrow \bar{A}$ ($\bar{\phi}$ the restriction of the multiplication ϕ). The submodule of primitives of a coalgebra A is the module $PA = \ker(\bar{A} \xrightarrow{\bar{\psi}} \bar{A} \otimes \bar{A})$ where $\bar{\psi}$ is induced by $\psi: A \to A \otimes A$.

A Hopf algebra is said to be primitively generated if the composition $PA \to A \to QA$ is surjective.

The Hopf-Borel theorem has an extension stating that if A is primitively generated, the x_i and y_j of 3.1.1 can be chosen to be primitive.

Recall that A is primitively generated if, and only if, the dual A^* is a commutative and associative algebra and for any $x \in A^*$ $x^p = 0$.

3.1.2. Remark: If E is a multiplicative theory so that $\alpha: E(X) \otimes E(X) \longrightarrow E(X \wedge X)$ is not an isomorphism $E(X)$ may fail to be a Hopf algebra (e.g. $E = H^*(\ ,Z)$). However, some of the notions used for Hopf algebras can be extended to generalized cohomology. e.g.: If E is represented by an Ω-spectrum $\{E_n, \gamma_n: E_n \to \Omega E_{n+1}\}$ (see 0.5) an element $x \in E^n(X)$ is called primitive if it can be prepresented by an H-map $X,\mu \to E_n$, Add.

An immediate consequence of the Hopf-Borel theorem is the following:

3.1.3. Proposition: *If* ΣX *is an H-space then* X *is an even dimensional sphere.*

Proof: In $H^*(\Sigma X, F)$ $\bar{\phi}: \bar{H}^*(\Sigma X, F) \otimes \bar{H}^*(\Sigma X, F) \longrightarrow \bar{H}^*(\Sigma X, F)$ is 0. By 3.1.1. the only possible Hopf algebra with this property is $\Lambda(x)$, and $\dim x = 2n+1$ if charac. $F \neq 2$. This holds for $F = Q$ or $F = Z_p$, hence $\Sigma X = S^{2n+1}$ and $X = S^{2n}$.

Actually a much stronger statement is true:

3.1.4. Theorem ([Adams]$_1$): *Among spheres only* S^1, S^3 *and* S^7 *admit H-structures.*

The various proofs of this theorem call for highly non-elementary techniques using either high order ordinary cohomology operations or primary operations in generalized cohomology theories.

This observation implies that the Hopf-Borel theorem, though an excellent first approximation, is far from being sufficient and more refined methods are desirable.

Some of these methods are described in the following sections.

3.2. Some relations between the algebra $H^*(X,Z_p)$ and the coalgebra $H^*(\Omega X, Z_p)$

Some of the relations in $H_*(\Omega X, Z_p)$ can be derived from $H^*(X, Z_p)$ via the Eilenberg-Moore spectral sequence where the arguments are purely algebraic. (See, e.g., [May and Zabrodsky]). We are about to give a more geometrical argument that by its nature may be applied to some generalized cohomology theories where the Eilenberg-Moore spectral sequence may fail to exist and may give some geometric insight to the algebraic arguments when the spectral sequence is available.

The following is somewhat related to 2.2.1.

3.2.1. Lemma: *Let* $f_0, F_0: X_0, \mu_0 \longrightarrow X_1, \mu_1$ *and* $f_1, F_1: X_1, \mu_1 \longrightarrow X_2, \mu_2$ *be H-maps. Let* $\xi_0: X_0 \to LX_1$, $\xi: X_0 \to LX_2$ *be homotopies* $* \sim f_0$ *and* $* \sim f_1 f_0$ *respectively.*

Define

$$\hat{w}_0: \quad X_0 \wedge X_0 \longrightarrow \Omega X_1$$

$$\hat{w}: \quad X_0 \wedge X_0 \longrightarrow \Omega X_2$$

$$v: \quad X_0 \longrightarrow \Omega X_2$$

by:

$$\hat{w}_0 \Lambda = w_0 = \xi_0 \mu_0 + F_0 - L\mu_1(\xi_0 \times \xi_0)$$

$$\hat{w}\Lambda = w = \xi\mu_0 + Pf_1F_0 + F_1(f_0 \times f_0) - L\mu_2(\xi \times \xi)$$

$$v = \xi - Lf_1\xi_0$$

Then $HD(v, \mu_0, \Omega\mu_2) = \hat{w} - \Omega f_1 \hat{w}_0$.

Proof:

$$w = (\xi\mu_0 - Lf_1\xi_0\mu_0) + (Lf_1\xi_0\mu_0 + Pf_1F_0 - Lf_1L\mu_1(\xi_0 \times \xi_0)) +$$
$$+ (Lf_1L\mu_1(\xi_0 \times \xi_0) + F_1(f_0 \times f_0) - L\mu_2(Lf_1 \times Lf_1)(\xi_0 \times \xi_0)) +$$
$$+ (L\mu_2(Lf_1\xi_0 \times Lf_1\xi_0) - L\mu_2(\xi \times \xi))$$

Now $PF_1(\xi_0 \times \xi_0)$ induces a homotopy $\phi: X_0 \times X_0 \longrightarrow L\Omega X_2$ $\phi: * \sim Lf_1L\mu_1(\xi_0 \times \xi_0) + F_1(f_0 \times f_0) - L\mu_2(Lf_1 \times Lf_1)(\xi_0 \times \xi_0)$.

As ΩX_2 is HC:

$$w = v\mu_0 + \Omega f_1 w_0 - \Omega\mu_2(v \times v) = v\mu_0 - \Omega\mu_2(v \times v) + \Omega f_1 w_0$$

$(\hat{w} - \Omega f_1\hat{w}_0)\Lambda = HD(v,\mu_0,\Omega\mu_2)\Lambda$ and Λ^* 1-1 implies 3.2.1.

$\hat{w}_0,\hat{w}$ of 3.2.1 are invariants defined whenever one has an H-map $f,F: X,\mu \to X',\mu'$ and a homotopy $\xi: X \to LX'$, $\xi: * \sim f$: $\hat{w}(F,\xi)\Lambda = \xi\mu + F - L\mu'(\xi \times \xi)$.

One can easily check that if $h: X \to X'$ and $\hat{\xi}: X \to LX'$ is a homotopy $* \sim h$ then if $\xi: \Omega X \to L\Omega X'$ is the obvious homotopy $\xi: * \sim \Omega h$ induced by $\hat{\xi}$ then for Ωh, $k(\Omega h)\mathrm{Add}: \Omega X, \mathrm{Add} \longrightarrow \Omega X', \mathrm{Add}$ $\hat{w}(k(\Omega h)\mathrm{Add}, \xi) = *$.

The nontriviality of $\hat{w}$ can be illustrated by:

3.2.2. Proposition: *For any space* X *let* $\bar{\Delta}: X \longrightarrow X \wedge X$ *be induced by the diagonal:* $\bar{\Delta} = \Lambda\Delta$. *Let* $\xi_0: \Omega X \to L\Omega(X \wedge X)$, $\xi_0: * \sim \Omega\bar{\Delta}$ *be the natural homotopy given by:* $\ell\xi_0 = \ell$

$$E_t\xi_0(\lambda) = \Omega\Lambda(\lambda + k_{\ell(\lambda)-t}(*),\ k_{\ell(\lambda)-t}(*) + \lambda)$$

Then $\hat{w}(k\Omega\bar{\Delta}\mathrm{Add}, \xi_0)T$ *is homotopic to* $u: \Omega X \wedge \Omega X \to \Omega^2(X \wedge X)$ *the adjoint of*

$$\Sigma^2(\Omega X \wedge \Omega X) = \Sigma\Omega X \wedge \Sigma\Omega X \xrightarrow{(1)_\# \wedge (1)_\#} X \wedge X.$$

(i.e.: $u(\lambda_1,\lambda_2)[s,t] = (\lambda_1(s), \lambda_2(t))$.)

<u>Proof</u>: The formal proof will be too complex to comprehend, thus useless. We shall give a less formal proof using the classical path spaces

$$\hat{P}X = X^I, \quad \hat{L}X = \{\gamma: I \to X \mid \gamma(0) = *\},$$

$$\hat{\Omega}X = \{\gamma: I \to X \mid \gamma(0) = * = \gamma(1)\}.$$

Now, it is quite difficult to describe $\hat{\Omega}^2 X \wedge X$. To overcome this difficulty consider

$$\tilde{\Lambda}: \hat{P}^2X \times \hat{P}^2X = \hat{P}^2(X \times X) = (X \times X)^{I^2} \longrightarrow \hat{P}^2(X \wedge X).$$

Then $Y_0 = \tilde{\Lambda}^{-1}[\hat{\Omega}^2(X \wedge X)]$ and $Y_1 = \tilde{\Lambda}^{-1}(\hat{L}\hat{\Omega}(X \wedge X))$ can be easily described and fortunately all maps and homotopies into $\hat{\Omega}(X \wedge X)$ involved in our proof factor through Y_0 and Y_1: $\hat{\Omega}\bar{\Delta}$ obviously factors through $PX \times PX$ and ξ_0 describes a map into Y_1. Now:

$$Y_0 = \{\gamma: I^2 \to X \times X \mid (p_1\gamma)^{-1}(*) \cup (p_2\gamma)^{-1}(*) \supset \partial I^2\}$$

$$Y_1 = \{\gamma: I^2 \to X \times X \mid (p_1\gamma)^{-1}(*) \cup (p_2\gamma)^{-1}(*) \supset J\},$$

$$J = \partial I \times I \cup I \times 0$$

We illustrate maps $\gamma: I^2 \to X \times X$ by marking on the domains I^2 of $p_1\gamma = \gamma_1$ and $p_2\gamma = \gamma_2$ the values they assume at various points. γ_i are constant along the marked lines and are constantly the base point $*$ along curled lines. Thus $\xi_0: \hat{\Omega}X \to \hat{L}\hat{\Omega}(X \wedge X)$ is marked by its representative $\hat{\Omega}X \to Y_1$ as follows:

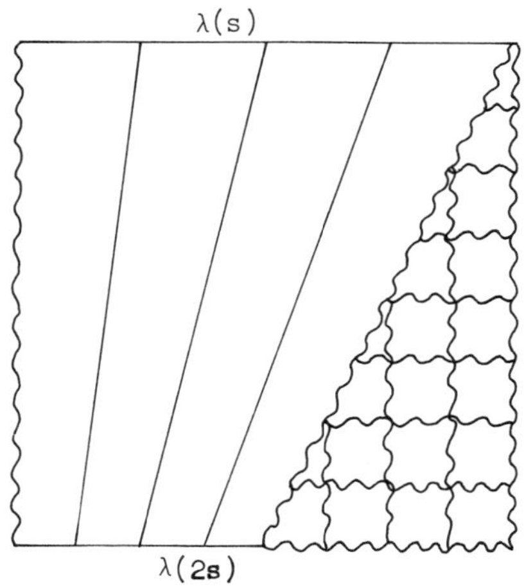

$p_1\xi_0(\lambda)$

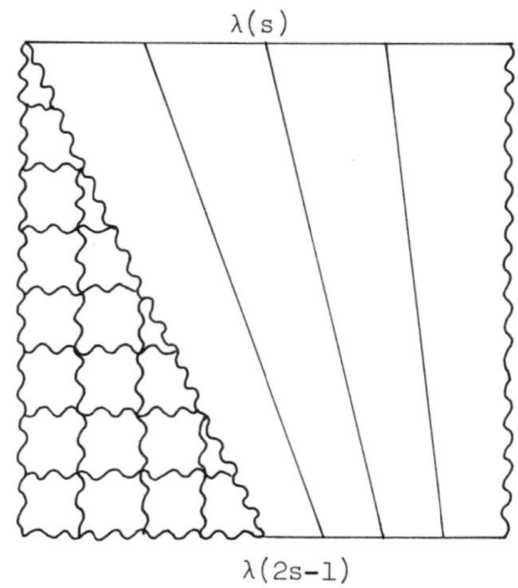

$p_2\xi_0(\lambda)$

$\xi_0(\lambda_1 + \lambda_2)$ is illustrated by

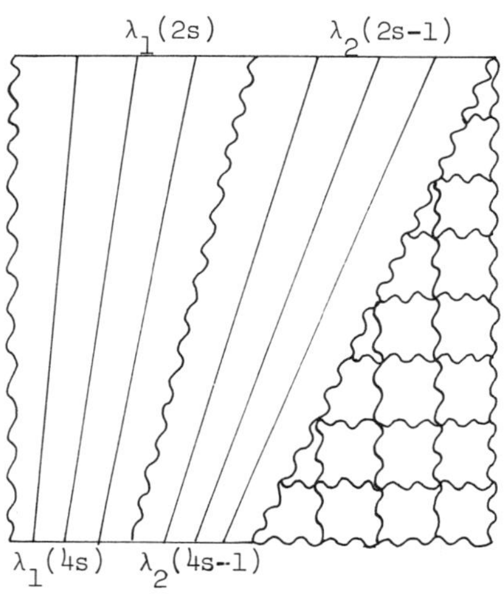

$p_1\xi_0(\lambda_1 + \lambda_2)$

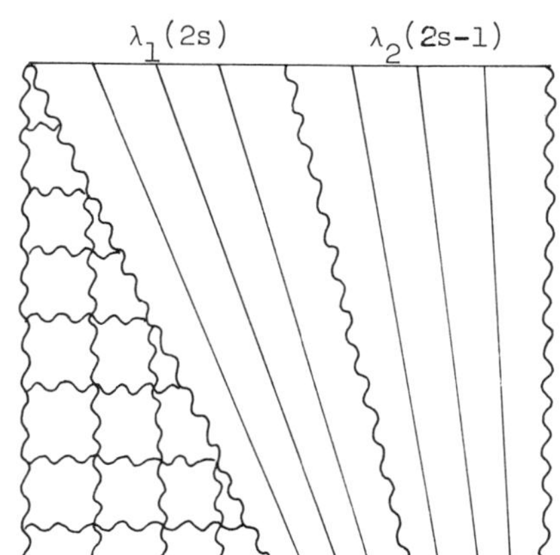

$p_2\xi_0(\lambda_1 + \lambda_2)$

$(L\ \mathrm{Add})(\xi_0(\lambda_1), \xi_0(\lambda_2))$ is given by:

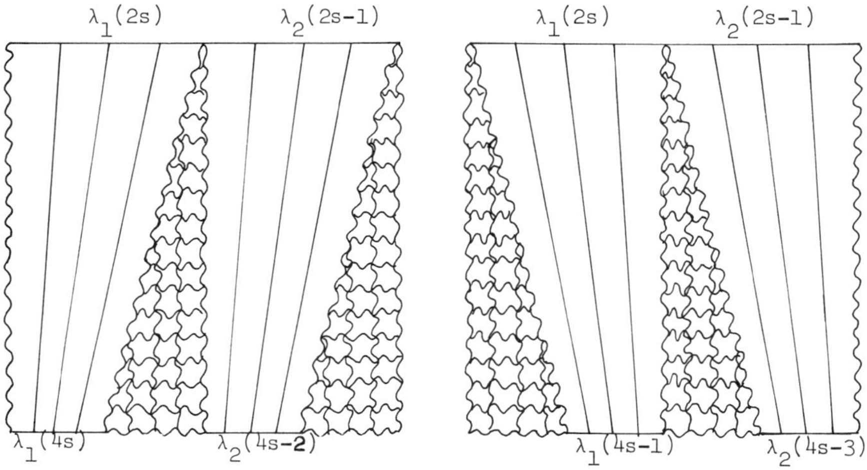

$p_1(L\ \mathrm{Add})(\xi_0(\lambda_1), \xi_0(\lambda_2))$ $p_2(L\ \mathrm{Add})(\xi_0(\lambda_1), \xi_0(\lambda_2))$

One can easily see that the images of all three maps $\hat{\Omega}X \times \hat{\Omega}X \longrightarrow \hat{P}^2(X \times X)$ lie in Y_1. The image of $\xi_0(\lambda_1 + \lambda_2) - (L\ \mathrm{Add})(\xi_0(\lambda_1), \xi_0(\lambda_2)) = w(\lambda_1, \lambda_2)$ illustrated below lies in Y_0:

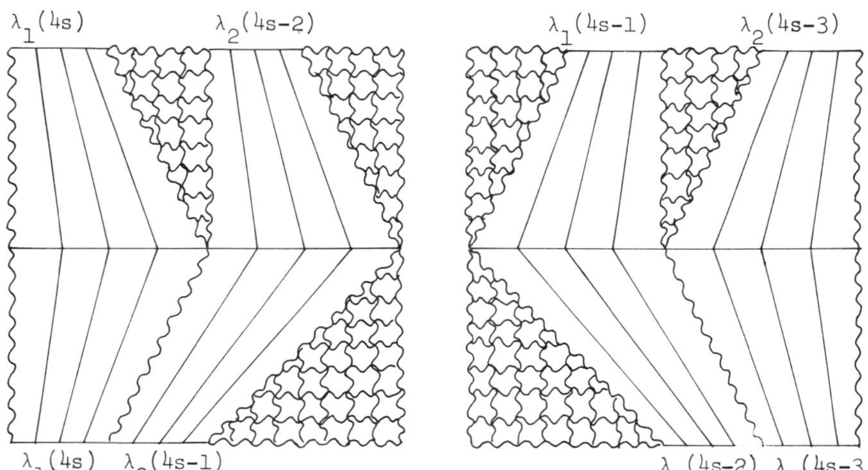

$p_1 w(\lambda_1, \lambda_2)$ $p_2 w(\lambda_1, \lambda_2)$

u is illustrated by

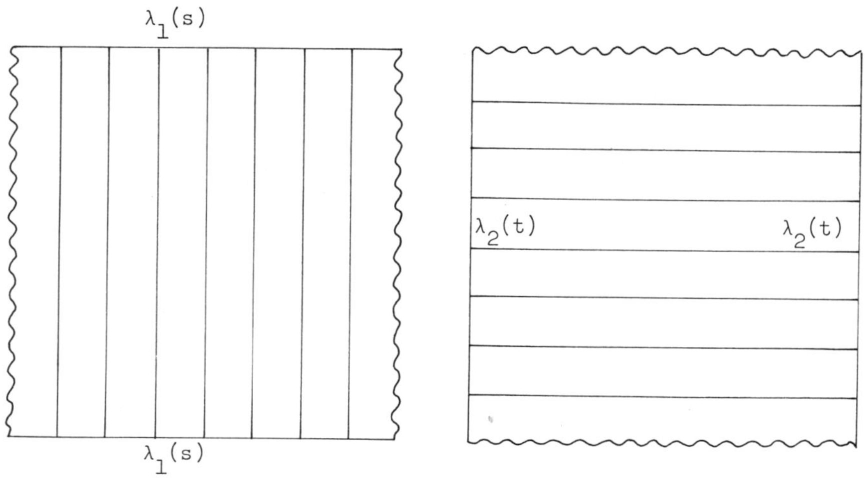

$p_1u(\lambda_1,\lambda_2)$ $\qquad\qquad$ $p_2u(\lambda_1,\lambda_2)$

One establishes the homotopy $wT \sim u\Lambda$ by deforming I^2 into itself via deformations ϕ_1^t, ϕ_2^t so that

$$p_1w(\lambda_1,\lambda_2)\phi_1^0 = p_1w(\lambda_1,\lambda_2), \quad p_2w(\lambda_1,\lambda_2)\phi_2^0 = p_2w(\lambda_1,\lambda_2),$$

$$p_1w(\lambda_1,\lambda_2)\phi_1^1 = p_1u(\lambda_2,\lambda_1) = \lambda_2p_1.$$

$$p_2w(\lambda_1,\lambda_2)\phi_2^1 = p_2u(\lambda_2,\lambda_1) = \lambda_1p_2 \quad \text{and}$$

$$(p_1w(\lambda_1,\lambda_2)\phi_1^t)^{-1}(*) \cup (p_2w(\lambda_1,\lambda_2)\phi_2^t)^{-1}(*) \supset \partial I$$

for every t.

ϕ_i^t deform I^2 in three stages:

(1) A deformation retraction followed by a deformation:

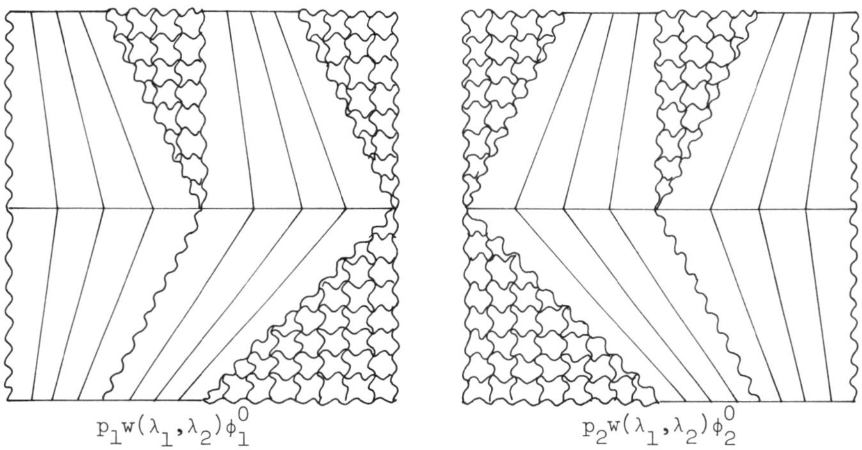

The (deformation) retraction:

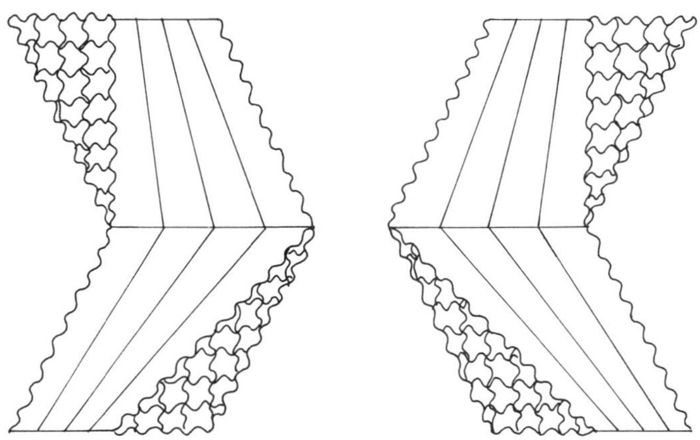

The deformation:

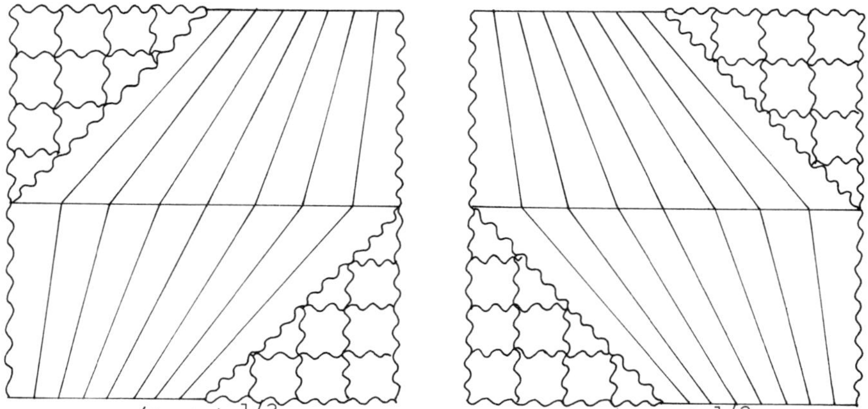

$p_1 w(\lambda_1,\lambda_2)\phi_1^{1/3}$ $p_2 w(\lambda_1,\lambda_2)\phi_2^{1/3}$

(2) A simple deformation

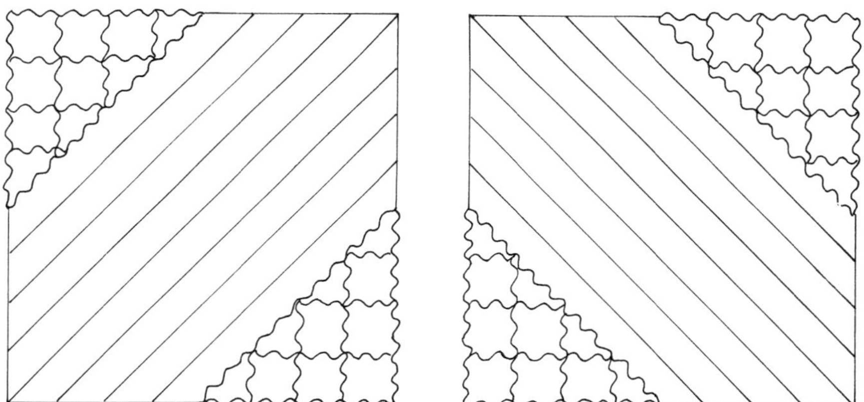

$p_1 w(\lambda_1,\lambda_2)\phi_1^{2/3}$ $p_2 w(\lambda_1,\lambda_2)\phi_2^{2/3}$

(3) A rotation:

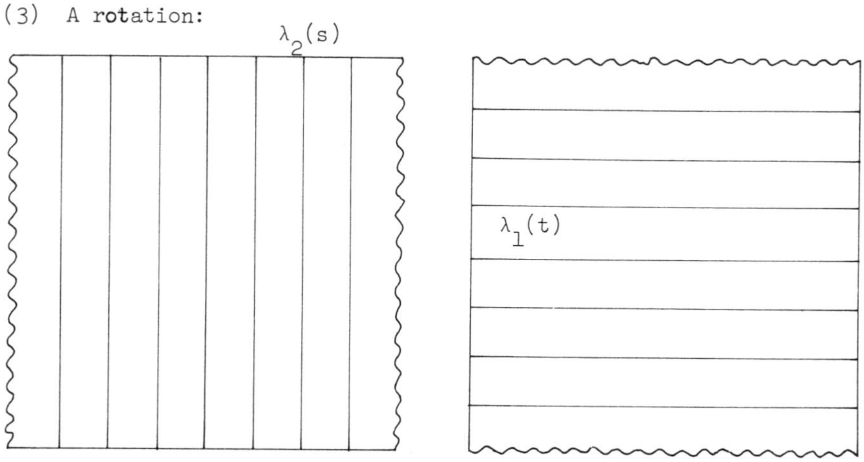

Our main result in this section can now be stated:

Let E^* be a cohomology theory represented by an Ω-spectrum $\{E_n,\gamma_n\}$ and a ring structure $\phi_{n,m}$ (see 0.5).

<u>3.2.3. Proposition</u>: *Given a space* X *and classes* $x_i \in E^{k_i}(X)$, $y_i \in E^{m_i}(X)$, $0 < k_i$, $0 < m_i$, $k_i + m_i = n$, $i = 1, 2,\ldots,a$. *If* $\sum_{i=1}^{a} x_i y_i = 0$ *then there exists* $z \in E^{n-2}(\Omega X)$ *so that* $\bar{\mu}^* z = \sum_{i=1}^{a} \alpha(\sigma y_i \otimes \sigma x_i)$ *where* $\bar{\mu}^*\colon E^*(\Omega X) \to E^*(\Omega X \wedge \Omega X)$ *is given by*

$$E^*(\Lambda)\bar{\mu}^* = E^*(\mathrm{Add}) - E^*(p_1) - E^*(p_2),$$

$\alpha\colon E^*(\Omega X) \otimes E^*(\Omega X) \to E^*(\Omega X \wedge \Omega X)$ *and* $\sigma\colon E^*(X) \to E^{*-1}(\Omega X)$ *as in* (0.5).

<u>Proof</u>: Let $x_i = [f_i] \in [X, E_{k_i}]$, $y_i = [g_i] \in [X, E_{m_i}]$. $\Sigma x_i y_i$ is then represented by the composition:

$$X \xrightarrow{\bar{\Delta}} X \wedge X \xrightarrow{\prod_{i=1}^{a} (f_i \wedge g_i)\Delta^a_{X \wedge X}} \prod_{i=1}^{a} (E_{k_i} \wedge E_{m_i}) \longrightarrow$$

$$\xrightarrow{\prod_{i=1}^{a} \phi_{k_i,m_i}} (E_n)^a \xrightarrow{(\mathrm{Add})^a} E_n$$

Put $h = (\mathrm{Add})^a(\prod_{i=1}^{a} \sigma_{k_i,m_i}) \prod_{i=1}^{a} (f_i \wedge g_i)\Delta^a_{X \wedge X}$. Then $\Sigma x_i y_i = 0$ is equivalent to $* \sim h\bar{\Delta}$. Let $\hat{\xi}\colon * \sim h\bar{\Delta}$. Apply 3.2.1 for

$$\Omega X \xrightarrow{\Omega\bar{\Delta}} \Omega(X \wedge X) \xrightarrow{\Omega h} \Omega E_n$$

i.e.: $X_0,\mu_0 = \Omega X$, Add, $X_1,\mu_1 = \Omega(X \wedge X)$, Add, $X_2,\mu_2 = \Omega E_n$, Add, $f_0 = \Omega\bar{\Delta}$, $f_1 = \Omega h$, F_0, F_1 - constant homotopies, ξ induced by $\hat{\xi}$ and

ξ_0 as in 3.2.2. Then, as was observed, $w = *$, by 3.2.2 $w_0 = uT$ and

$$-Hd(v,Add,Add) = (\Omega^2 h)w_0 = \Omega^2 huT$$

One has the following commutative diagram

$$\begin{array}{ccccc}
\Omega X \wedge \Omega X & \xrightarrow{\quad u \quad} & \Omega^2(X \wedge X) & \xrightarrow{\quad \Omega^2 h \quad} & \\
\Big\downarrow {\scriptstyle [\prod_{i=1}^{a} (\Omega f_i \wedge \Omega g_i)]\Delta^a_{\Omega X \wedge \Omega X}} & & \Big\downarrow {\scriptstyle \Omega^2[\prod_{i=1}^{a} f_i \wedge g_i]\Delta^a_{\Omega^2(X \wedge X)}} & & \Big\downarrow \\
\prod_{i=1}^{a} (\Omega E_{m_i} \wedge \Omega E_{k_i}) & \longrightarrow & \prod_{i=1}^{a} \Omega^2(E_{m_i} \wedge E_{k_i}) & \xrightarrow{\Omega^2\{(Add)^a \prod_{i=1}^{a} \phi_{m_i,k_i}\}} & \Omega^2 E_n
\end{array}$$

Hence, $-HD(w,Add,Add) = \sum_{i=1}^{a} \alpha(\sigma y_i \otimes \sigma x_i)$ and $-[HD(w,Add,Add) = \bar{\mu}^*[w]$. Put $-[w] = z$.

3.2.4. Example (see [May and Zabrodsky] for another approach):

The Hopf algebra $H^*(Spin(n), Z_p)$:

One has a fibration $Spin(n) \xrightarrow{r} SO(n) \xrightarrow{k} K(Z_2,1)$ and one can apply the Eilenberg-Moore s.s.: $H^*(SO(n), Z_2) = \prod_{\substack{m < n \\ m=odd}} Z_2[x_m]/(x_m^{2^{\alpha(m)}})$ where $\alpha(m)$ is given by $m2^{\alpha(m)} \geq n > m2^{\alpha(m)-1}$. $H^*(k,Z_2)\iota_1 = x_1$. $H^*(K(Z_2,1), Z_2) = Z_2[\iota_1]$ and $Z_2[\iota_1] \approx Z_2[\iota_1^{2^{\alpha(1)}}] \otimes Z_2[\iota_1]/(\iota_1^{2^{\alpha(1)}})$ as a $Z_2[\iota_1^{2^{\alpha(1)}}]$ module.

$$\mathrm{Tor}^{H^*(K(Z_2,1),Z_2)}(H^*(SO(n),Z_2),Z_2) \approx$$

$$\approx H^*(SO(n),Z_2)/(x_1) \otimes \mathrm{Tor}^{Z_2[\iota_1^{2^{\alpha(1)}}]}(Z_2,Z_2) \approx$$

$$\approx H^*(SO(n),Z_2)/(x_1) \otimes \Lambda(x_{2^{\alpha(1)}-1})$$

$E_2^{0,*} = H^*(SO(n),Z_2)/(x_1)$, $x_{2^{\alpha(1)}-1} \in E_2^{1,*}$ and obviously $E_2 = E_\infty = H^*(\mathrm{Spin}(n),Z_2)$ as an algebra.

The co-algebra of $H^*(\mathrm{Spin}(n),Z_2)$ will be determined by $H^*(\mu,Z_2)x_{2^{\alpha(1)}-1}$ as $H^*(SO(n),Z_2) \to H^*(SO(n),Z_2)/(x_1) \subset H^*(\mathrm{Spin}(n),Z_2)$ is a map of Hopf algebras. Dually, one has an exact sequence of Hopf algebras $0 \to \Lambda(\hat{x}_{2^{\alpha(m)}-1}) \to H_*(\mathrm{Spin}(n),Z_2) \to \mathrm{im}\, H_*(r,Z_2) = \bigotimes_{\substack{i<n \\ i\neq 2^k}} \Lambda(\hat{x}_i) \to 0.$

If $H_*(\mathrm{Spin}(n),Z_2)$ is commutative (and it is obviously associative), $H^*(\mathrm{Spin}(n),Z_2)$ is primitively generated.

As the lowest dimensional commutator of $H_*(\mathrm{Spin}(n),Z_2)$ is primitive, the lowest dimension where a commutator can occur is $2^{\alpha(1)}-1$ (as in dim $< 2^{\alpha(1)}-1$ $H_*(\mathrm{Spin}(n), Z_2) \to H_*(SO(n),Z_2)$ is surjective) and it has to involve a commutator of two generators, hence it involves $[\hat{x}_i,\hat{x}_j]$ $i+j = 2^{\alpha(1)}-1$. If $n-1 = 2^{\alpha(1)-1}$ the relations $i < n$, $j < n$, $i \neq 2^k \neq j$, $i+j = 2^{\alpha(1)}-1$, is impossible and $H^*(\mathrm{Spin}(n),Z_2)$ is primitively generated.

If $n-1 > 2^{\alpha(1)-1}$ we shall show that $[\hat{x}_i,\hat{x}_j] = \hat{x}_{2^{\alpha(1)}-1} \neq 0$ for all i,j, $i+j = 2^{\alpha(1)}-1$ or dually: $\bar{\mu}^*x_{2^{\alpha(1)}-1} = \sum_{\substack{i+j=2^{\alpha(1)}-1 \\ i<j}} x_i \otimes x_j.$

To see that, consider the fibration

$$BSpin(n) \xrightarrow{Br} BSO(n) \xrightarrow{Bk} K(Z_2,2),$$

$$H^*(Bk,Z_2)\iota_2 = (Bk)^*\iota_2 = w_2 \in \ker(Br)^*.$$

Now, $H^*(BSO(n),Z_2) = Z_2\ w_2, w_3,\ldots,w_n],\ \ n \leq 2^{\alpha(1)}$.

$$v_{2^{\alpha(1)}+1} = Sq^{2^{\alpha(1)-1}} Sq^{2^{\alpha(1)-2}} \ldots Sq^4 Sq^2 Sq^1 w_2 \in H^{2^{\alpha(1)}+1}(BSO(n),Z_2)$$

is decomposable and in $\ker(Br)^*$. By the well-known Wu formulas, if $n \neq 2^{\alpha(1)-1}+1$ $\ v_{2^{\alpha(1)}+1} = \sum_{i+j=2^{\alpha(1)}+1} w_j \cdot w_i + \Sigma w_i d_i$, d_i decomposables hence in $\ker\sigma$. $H^*(Br,Z_2)v_{2^{\alpha(1)}+1} = \Sigma Br^*w_j \cdot Br^*w_i + \Sigma Br^*w_i \cdot Br^*d_i = 0$. By 3.2.3 there exists $z \in H^{2^{\alpha(1)}-1}(Spin(n),Z_2)$, $\bar{\mu}^*z = \sum_{i+j=2^{\alpha(1)}+1} \sigma Br^*w_j \otimes \sigma Br^*w_i$. $\sigma Br^*w_j = x_{j-1}$, $Br^*w_i = x_{i-1}$ and one may assume $z = x_{2^{\alpha(1)}-1}$.

3.3. Browder's Bockstein Spectral Sequence

Given an Ω-spectrum $\{E_n,\ \gamma_n\colon E_n \xrightarrow{\approx} \Omega E_{n+1}\}$ let $\lambda_p = \lambda_{p,n}\colon E_n \to E_n$ be induced by the p-th power map with respect to loop multiplication. Let $\delta\colon E_{n-1}(Z_p) \to E_n$ be the fiber of $\lambda_{p,n}$. Then one gets a Puppe sequence of fibrations - the Z_p coefficient sequence:

$$(3.3.1) \dashrightarrow E_n \xrightarrow{\lambda_{p,n}} E_n \xrightarrow{\rho} E_n(Z_p) \xrightarrow{\delta} E_{n+1} \xrightarrow{\lambda_{p,n+1}} E_{n+1} \dashrightarrow$$

This sequence gives rise to the following natural exact sequence:

$$(3.3.2) \dashrightarrow E^n(X) \xrightarrow{(\lambda_p)_*} E^n(X) \xrightarrow{\rho_*} E^n(X,Z_p) \xrightarrow{\delta_*} E^{n+1}(X) \xrightarrow{(\lambda_p)_*} \cdots$$

As $(\lambda_p)_*$ is multiplication by p one obtains a "universal coefficients" exact sequence.

(3.3.3) $$0 \to E^n(X) \otimes Z_p \xrightarrow{\rho'_*} E^n(X,Z_p) \to \mathrm{Tor}(E^{n+1}(X),Z_p) \to 0$$

as well as an exact couple

(3.3.4) $$\begin{array}{ccc} E^*(X) & \xrightarrow{(\lambda_p)_*} & E^*(X) \\ {\scriptstyle \delta_*}\nwarrow & & \swarrow{\scriptstyle \rho_*} \\ & E^*(X,Z_p) & \end{array}$$

The spectral sequence induced by (3.3.4) first studied by W. Browder is called the Bockstein spectral sequence.

For simplicity assume that $E^n(X)$ is finitely generated (e.g.: suppose $\pi_*(E)$ is finitely generated, in particular, $\pi_m(E) = 0$ for $m > N(E)$).

The main properties of the Bockstein spectral sequence are given by

<u>3.3.5. Theorem</u>: 3.3.4 *yields a spectral sequence* $\{E_r, \beta_r\}$ *so that*

(a) $E_1 = E^*(X,Z_p)$, $\beta_1 = \beta_*: E^n(X,Z_p) \to E^{n+1}(X,Z_p)$ *given by the exact sequence*

(3.3.5.1) $$--- \to E^n(X,Z_{p^2}) \xrightarrow{\hat{\rho}_*} E^n(X,Z_p) \xrightarrow{\beta_*} E^{n+1}(X,Z_p) \to E^{n+1}(X,Z_{p^2})$$

induced by the top line fibration in

(3.3.5.2)

$$\begin{array}{ccccccc}
E_n(Z_p) & \longrightarrow & E_n(Z_{p^2}) & \longrightarrow & E_n(Z_p) & \xrightarrow{\beta} & E_{n+1}(Z_p) \\
\| & & \Big\downarrow \delta_2 & & \Big\downarrow \delta & & \| \\
E_n(Z_p) & \longrightarrow & E_{n+1} & \xrightarrow{\lambda_p} & E_{n+1} & \xrightarrow{\rho} & E_{n+1}(Z_p) \\
 & & \Big\downarrow \lambda_{p^2} & & \Big\downarrow \lambda_p & & \\
 & & E_{n+1} & = & E_{n+1} & &
\end{array}$$

(The derivations β_r *are called the higher Bocksteins.)*

(b) $E_\infty^n = \mathrm{im}\{[E^n(X)/\mathrm{torsion}] \otimes Z_p \to E^n(X,Z_p)/\rho^!_*(\mathrm{torsion}\ E^n(X) \otimes Z_p)\}$

$\approx [E^n(X)/\mathrm{torsion}] \otimes Z_p.$

Proof: (a) By the general theory of spectral sequences (e.g.: [Hu], p. 232) $E_1^* = E^*(X,Z_p)$ and $\beta_1 = \rho_*\delta_* = (\rho\delta)_* = \beta_*$.

(b) $E_\infty = \delta_*^{-1}(\bigcap_{n=1}^{\infty} \mathrm{im}(\lambda_p)_*^n)/\rho_*[\bigcup_{n=1}^{\infty} \ker(\lambda_p)_*^n]$.

As $E^n(X)$ is a finitely generated abelian group $\bigcap_{n=1}^{\infty} \mathrm{im}(\lambda_p)_*^n = \neq p$ torsion of $E^n(X)$ and $\bigcup_{n=1}^{\infty} \ker(\lambda_p)_*^n = p$ torsion of $E^n(X)$. $E^*(X,Z_p)$ is a p-torsion group so by 3.3.2 and 3.3.3 one has $\delta_*^{-1}(\neq p$ torsion of $E^n(X)) = \ker \delta_* = \mathrm{im}\, \rho_* = \mathrm{im}\, \rho^!_*$.

Hence,

$$E_\infty = \rho^!_*(E^n(X) \otimes Z_p)/\rho^!_*[(p\text{-torsion } E^n(X)) \otimes Z_p]$$

$$= \mathrm{im}[(E^n(X)/\mathrm{torsion}) \otimes Z_p \to E^n(X,Z_p)/\rho^!_*(\mathrm{torsion}\ E^n(X) \otimes Z_p)]$$

and by

$$\begin{array}{ccccc}
 & & 0 & & \\
 & & \downarrow & & \\
 & & [\text{torsion } E^n(X)] \otimes Z_p & = & [\text{torsion } E^n(X)] \otimes Z_p \\
 & & \downarrow & & \downarrow \\
0 & \longrightarrow & E^n(X) \otimes Z_p & \xrightarrow{\rho'_*} & E^n(X, Z_p) \\
 & & \downarrow & & \downarrow \\
 & & [E^n(X)/\text{torsion}] \otimes Z_p & \xrightarrow{\hat{\rho}_*} & E^n(X,Z_p)/\rho'_* \text{ torsion } E^n(X) \otimes Z_p \\
 & & \downarrow & & \downarrow \\
 & & 0 & & 0
\end{array}$$

$\hat{\rho}_*$ is an injection and $E^n_\infty = [E^n(X)/\text{torsion}] \otimes Z_p$.

Note that $E^n(X,Z_p)$ is a Z_p vector space only for p-odd. (For $p = 2$ take E to be the 2-stage Postnikov system with Sq^2 as a k-invariant and let $X = M(Z_2,n)$ the Moore space. Then $E^n(X) = Z_4$.)

However, the above spectral sequence holds for $p = 2$ and E_∞ is then a Z_2 vector space.

One obvious application of 3.3.5 is the following:

<u>3.3.6. Corollary</u>: $E^*(X)$ *is* p-*torsion free if and only if* $E^*(X,Z_p)$ *is a* Z_p*-vector space (redundant for* $p \neq 2$*) and rank* $E^n(X,Z_p) = \text{rank } E^n(X) = \text{rank } (E^n(X) \otimes Q)$ *for every* n.

<u>Proof</u>: If $E^*(X)$ is p-torsion free then by (3.2.3) $E^n(X,Z_p) \approx E^n(X) \otimes Z_p \approx (E^n(X)/\text{torsion}) \otimes Z_p$ and for any p-torsion free abelian group G rank G = rank $(G \otimes Z_p)$.

Conversely, if $E^*(X,Z_p)$ is a Z_p vector space with rank $E^*(X,Z_p)$ = rank $E^*(X)$ then rank E_1 = rank E_∞ in the Bockstein spectral sequence, $E_1 = E_\infty$, $E^*(X,Z_p) \approx E^*(X,Z_p)/\rho_*[(\text{torsion } E^*(X)) \otimes Z_p]$, $[\text{torsion } E^*(X)] \otimes Z_p = 0$ and $E^*(X)$ is p-torsion free.

The following is a geometric interpretation of the Bockstein spectral sequence:

<u>3.3.7. Lemma</u>: *Let* X *be a space,* E^* *a cohomology theory given by an* Ω-spectrum $\{E_n, \gamma_n : E_n \xrightarrow{\approx} \Omega E_{n+1}\}$. *Then* $x \in E^n(X,Z_p) = [X, E_n(Z_p)]$ *survives to* E_r *in the Bockstein s.s. if and only if the representation* $f_1 : X \to E_n(Z_p)$ *of* x *can be lifted to* $f_r : X \to E_n(Z_{p^r})$, $\rho_r f_r \sim f_1$. $\beta_r\{x\} \in E_r$ *is then represented by* $\hat{\beta}_r f_r$, $\hat{\beta}_r : E_n(Z_{p^r}) \to E_{n+1}(Z_p)$. ρ_r *and* $\hat{\beta}_r$ *are given by the following diagram:*

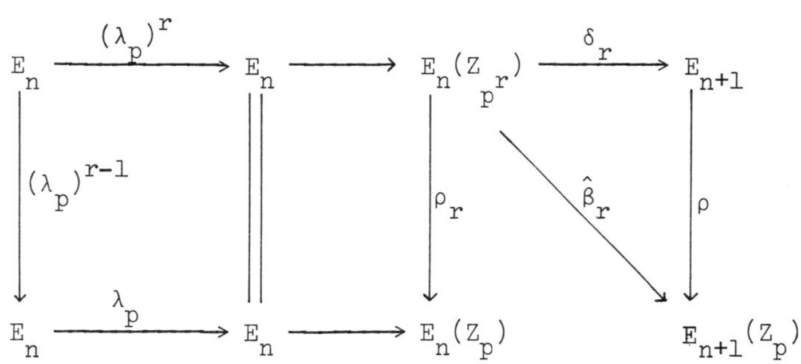

<u>Proof</u>: By the general theory of spectral sequences

$$E_r = \delta_*^{-1}(\text{im}(\lambda_p)_*^{r-1})/\rho_* \ker(\lambda_p)_*^{r-1}$$

hence, $x \in E_1$ survives to E_r if $\delta_* x \in \text{im}(\lambda_p)_*^{r-1}$ or, if δf_1 lifts to $\hat{f}_r$, $\lambda_p^{r-1}\hat{f}_r = \delta f_1$. In the following diagram $E_n(Z_{p^r})$, δ_r and ρ_r are the pull back of δ and λ_p^{r-1}, $\delta_r f_r = \hat{f}_r$.

$$\begin{array}{ccccccc} & & E_n(Z_{p^r}) & \xrightarrow{\delta_r} & E_{n+1} & \xrightarrow{\rho} & E_{n+1}(Z_p) \\ & \nearrow^{f_r} & \downarrow{\rho_r} & & \downarrow{\lambda_p^{r-1}} & & \\ X & \xrightarrow{f_1} & E_n(Z_p) & \xrightarrow{\delta} & E_{n+1} & & \end{array}$$

By definition of β_r $\beta_r\{x\}$ is represented by

$$\rho\hat{f}_r = \rho\delta_r f_r = (\rho\delta_r)_* f_r = (\hat{\beta}_r)_* f_r.$$

In the remainder of this section we restrict our study to $E_n = K(Z,n)$. In this case the Bockstein spectral sequence for an H-space is a spectral sequence of Hopf algebras. Here β is the ordinary (primary) Bockstein operator and it is well known that β is a derivation: $\beta^2 = 0$, $\beta(xy) = (\beta x)y + (-1)^{|x|}x\beta y$.

Let $\widetilde{\beta}_r\,\iota \in H^{n+1}(K(Z_{p^r},n),\ Z_{p^r}) = Z_{p^r}$ be a generator. ($\widetilde{\beta}_r$ is the fiber of the map $K(Z_{p^r},\ n+1) \to K(Z_{p^{2r}},\ n+1)$ induced by the inclusion $Z_{p^r} = p^r Z_{p^{2r}} \subset Z_{p^{2r}}$.) Then, $\beta_1 = \beta$, $\rho_r\widetilde{\beta}_r = \hat{\beta}_r$ and $\widetilde{\beta}_r$ is a derivation.

For convenience in the sequel, suppose p is odd. One has the following:

<u>3.3.8. Proposition</u>: *Let* $x \in E_r^{2n}(X)$ *in the Bockstein s.s. of a space* X, $r > 1$. *Then* x^p *and* $x^{p-1}\beta_r x$ *survive to* $E_{r+1}(X)$ *and* $\beta_{r+1}\{x^p\} = \{x^{p-1}\beta_r x\}$.

<u>Proof</u>: $\beta_r x^p = px^{p-1}\beta_r x = 0$, $\beta_r(x^{p-1}\beta_r x) = (p-1)x^{p-2}(\beta_r x)^2 = 0$ as $\dim \beta_r x$ is odd and thus $(\beta_r x)^2 = 0$.

By the naturality of the spectral sequence and by 3.3.7 suffices to consider the case $X = K(Z_{p^r}, 2n)$, $x = \{\hat{\iota}_{2n}\}$ where $\hat{\iota}_{2n} \in H^{2n}(K(Z_{p^r}, 2n), Z_p) = Z_p$ is the reduction of the fundamental class $\tilde{\iota}_{2n}$. Consider the following:

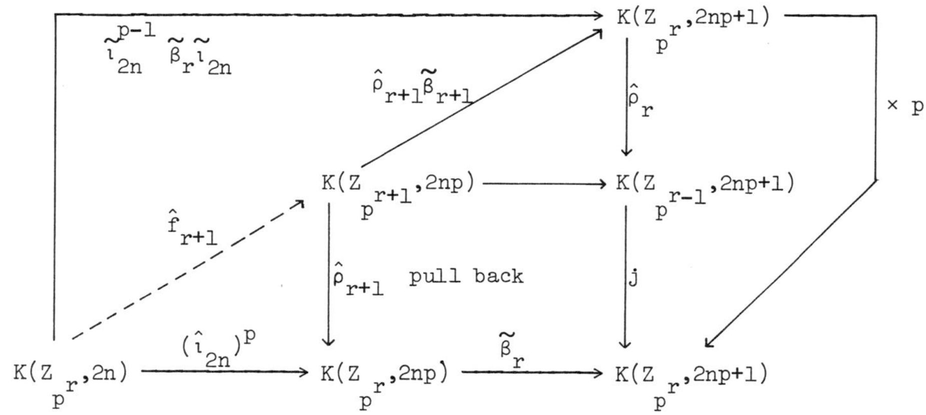

$\tilde{\beta}_r\tilde{\iota}^p_{2n} = p(\tilde{\iota}^{p-1}_{2n}\tilde{\beta}_r\tilde{\iota}_{2n})$, hence the diagram (excluding $\hat{f}_{r+1}$) is commutative. $\hat{f}_{r+1}$ exists to satisfy $\hat{\rho}_r\hat{\rho}_{r+1}\tilde{\beta}_{r+1}\hat{f}_{r+1} = \hat{\rho}_r(\tilde{\iota}^{p-1}_{2n}\tilde{\beta}_r\hat{\iota}_{2n})$.

By 3.3.7, $\beta_{r+1}\{\hat{\iota}^p_{2n}\} = \rho_r\hat{\rho}_{r+1}\tilde{\beta}_{r+1}\hat{f}_{r+1} = \rho_{r-1}\hat{\rho}_r\hat{\rho}_{r+1}\tilde{\beta}_{r+1}\hat{f}_{r+1} =$
$= \rho_r(\tilde{\iota}^{p-1}_{2n}\tilde{\beta}_r\tilde{\iota}_{2n}) = \{\hat{\iota}^{p-1}_{2n}\beta_r\hat{\iota}_{2n}\}$.

For $r = 1$ the proof of 3.3.8 needs more care. For this and for the next section we need the following lemma (stated for p odd but has an analogue for $p = 2$).

3.3.9. A fundamental lemma: *Let* $\xi: K(Z_p,2n) \to K(Z_p, 2np)$ *be given by* $H^*(\xi,Z_p)\iota_{2np} = \iota^p_{2n}$. *Then* ξ *lifts to* $\hat{\xi}: K(Z_p,2n) \to K(Z_{p^2}, 2np)$ *and for any such map* $HD(\hat{\xi}) = HD(\hat{\xi},Add,Add) = j(v + \overline{\mu}^*d)$.

$$v = \sum_{k=1}^{p-1} \frac{1}{p}\binom{p}{k} \ \iota_{2n}^{k} \otimes \iota_{2n}^{p-k}$$

where $j\colon K(Z_p,\ 2np) \to K(Z_{p^2},\ 2np)$ *is induced by the obvious injection* $Z_p \to Z_{p^2}$ *and* $\overline{\mu}^* = H^*(\mu, Z_p) - H^*(p_1, Z_p) - H^*(p_2, Z_p)$.

Consequently, if $K(Z_p,\ 2np+1) \xrightarrow{\hat{j}} E \xrightarrow{r} K(Z_p,\ 2n+1)$ *is the fibration induced by* $\beta P^n\colon K(Z_p,\ 2n+1) \longrightarrow K(Z_p,\ 2np+2)$ *then there exists* $\chi\colon K(Z_p,2n) \longrightarrow \Omega E$, $\Omega r \chi = 1$ *so that*

$$\mathrm{Add}(\chi \times \Omega\hat{j})\colon K(Z_p,2n) \times K(Z_p,2np),\ \ \mu_v \longrightarrow \Omega E, \mathrm{Add}$$

is an H-homotopy equivalence, $v\colon K(Z_p,2n) \wedge K(Z_p,2n) \longrightarrow K(Z_p,2np)$ *as above and* $\mu_v[(x,y),\ (\overline{x},\overline{y})] = x\cdot\overline{x},\ \ v(x,\overline{x})\cdot y\cdot\overline{y}$.

<u>Proof</u>: We first show that no lifting $\hat{\xi}$ of ξ is an H-map by showing that $HD(\hat{\xi},\mathrm{Add},\mathrm{Add}) = j(v + \overline{\mu}^*d + z_\beta)$ where z_β is in I_β - the $A(p)$ ideal in $H^*(K(Z_p,2n),\ Z_p) \otimes H^*(K(Z_p,2n),\ Z_p)$ generated by all classes of the form $\beta a \iota_{2n} \otimes \iota_{2n}$ and $\iota_{2n} \otimes \beta a \iota_{2n}$, $a \in A(p)$.

By 2.2 type argument one can see that altering the lifting $\hat{\xi}$ of ξ by an element $d_1 \in [K(Z_p,2n),\ K(Z_p,2np)]$ $HD(\hat{\xi})$ will be altered by $j\overline{\mu}^*d_1$ and so one can prove the above formula for any particular choice of $\hat{\xi}$.

One has the following diagram:

$$\begin{array}{ccc}
K(Z,2n) & \xrightarrow{(\tilde{\iota}_{2n})^p} & K(Z,2np) \\
\downarrow{\scriptstyle\rho} & & \downarrow{\scriptstyle\rho} \\
K(Z_p,2n) & \xrightarrow{\ \ \xi\ \ } & K(Z_p,2np)
\end{array}$$

As $\beta\iota^{p}_{2n} = 0$ ξ lifts to $\hat{\xi}\colon K(Z_p,2n) \longrightarrow K(Z_{p^2}, 2np)$. As ρ factors as $K(Z,2np) \xrightarrow{\rho_2} K(Z_{p^2},2np) \xrightarrow{\hat{\rho}_1} K(Z_p,2np)$ one obtains

$$\begin{array}{ccc} K(Z,2n) & \xrightarrow{(\tilde{\iota}_{2n})^p} & K(Z,2np) \\ \Big\downarrow \rho & & \Big\downarrow \rho_2 \\ K(Z_p,2n) & \xrightarrow{\hat{\xi}} & K(Z_{p^2}, 2np) \end{array}$$

which need not be commutative for an arbitrary $\hat{\xi}$. However, $\hat{\rho}_1\hat{\xi}\rho = \hat{\rho}_1\rho_2(\tilde{\iota}_{2n})^p$ implies the existence of $\tilde{d}_1\colon K(Z,2n) \longrightarrow K(Z_p, 2np)$ so that $\rho_2(\tilde{\iota}^{p}_{2n}) = j\tilde{d}_1 + \hat{\xi}\rho$. Now, $H^*(K(Z_p,2n), Z_p) \to H^*(K(Z,2n), Z_p)$ is surjective ($\ker H^*(\rho,Z_p)$ is the $A(p)$ ideal generated by $\beta\iota_{2n}$), hence $\tilde{d}_1 = d_1\rho$ and altering $\hat{\xi}$ by jd_1 if necessary one may assume that $\hat{\xi}\rho = \rho_2(\tilde{\iota}_{2n})^p$. As ρ and ρ_2 are H-maps

$$HD(\hat{\xi})(\rho \wedge \rho) = HD(\hat{\xi}\rho) = HD(\rho_2(\tilde{\iota}_{2n})^p) = \rho_2 HD(\tilde{\iota}^{p}_{2n})$$

If $p_i\colon K(Z,2n) \times K(Z,2n) \longrightarrow K(Z,2n)$ $i = 1,2$ are the projections then $H^*(Add,Z_p)\tilde{\iota}_{2n} = H^*(p_1,Z_p)\tilde{\iota}_{2n} + H^*(p_2,Z_p)\tilde{\iota}_{2n} = p_1^*\tilde{\iota}_{2n} + p_2^*\tilde{\iota}_{2n}$.

$$\begin{aligned} HD(\tilde{\iota}^{p}_{2n}) &= H^*(Add,Z_p)(\tilde{\iota}_{2n})^p - p_1^*(\tilde{\iota}_{2n})^p - p_2^*(\tilde{\iota}^{p}_{2n}) = \\ &= (p_1^*\tilde{\iota}_{2n} + p_2^*\tilde{\iota}_{2n})^p - [p_1^*(\tilde{\iota}_{2n})]^p - [p_2^*(\tilde{\iota}_{2n})]^p = \\ &= p\ \Sigma\frac{1}{p}\binom{p}{k}(p_1^*\tilde{\iota}_{2n})^k \cdot (p_2^*\tilde{\iota}_{2n})^{p-k} = p\tilde{v}_0. \end{aligned}$$

Using the commutativity of

$$\begin{array}{ccc} K(Z,2np) & \xrightarrow{\ \rho\ } & K(Z_p,2np) \\ \big\downarrow{\scriptstyle \times p} & & \big\downarrow{\scriptstyle j} \\ K(Z,2np) & \xrightarrow{\ \rho_2\ } & K(Z_{p^2},2np) \end{array}$$

$HD(\hat{\xi})(\rho \wedge \rho) = \rho_2 HD(\widetilde{\iota}^{p}_{2n}) = j\rho\widetilde{v}_0 = jv(\rho \wedge \rho)$

Now, the fact that ξ is an H-map implies that $HD(\hat{\xi}) = j\hat{w}$, $HD(\hat{\xi})\rho \wedge \rho = j\hat{w}(\rho \wedge \rho) = j\rho\widetilde{v}_0$. Again, $j[\hat{w}(\rho \wedge \rho) - \rho\widetilde{v}_0] = 0$ implies that $\hat{w}(\rho \wedge \rho) - \rho\widetilde{v}_0$ factors through fiber j which is β: $K(Z_p, 2np-1) \longrightarrow K(Z_p, 2np)$ and by the surjectivity of $H^*(\rho \wedge \rho, Z_p)$ $\hat{w}(\rho \wedge \rho) - \rho\widetilde{v}_0 = \beta\hat{w}_1(\rho \wedge \rho)$ and altering $\hat{w}$ by $\beta\hat{w}_1$ if necessary one has $\hat{w}(\rho \wedge \rho) = \rho\widetilde{v}_0 = v(\rho \wedge \rho)$, $\hat{w} = v + z_\beta$, $z_\beta \in I_\beta \supset \ker H^*(\rho \wedge \rho, Z_p)$.

To show that for any choice of $\hat{\xi}$ $HD(\hat{\xi}) \neq 0$ one has to show the impossibility of $-v = \overline{\mu}^*d + z_\beta$.

Let $(\overline{\mu}^*)^m$: $\overline{H}^*(K(Z_p,2n), Z_p) \to \overline{H}(K(Z_p,2n), Z_p))^m$

$$= \underbrace{\overline{H}^*(K(Z_p,2n), Z_p) \otimes \overline{H}^*(K(Z_p,2n), Z_p) \otimes \ldots \otimes H^*(K(Z_p,2n),Z_p)}_{m}$$

$(\overline{\mu}^*)^m$ given by $(\overline{\mu}^*)^2 = \overline{\mu}^*$, $(\overline{\mu}^*)^m = [(\overline{\mu}^*)^{m-1} \otimes 1]\overline{\mu}^*$.

Let $(\overline{\mu}_*)^m$ be the dual of $(\overline{\mu}^*)^m$, then $(\overline{\mu}_*)^m\underbrace{(x \otimes x \ldots \otimes x)}_{m} = x^m$.

One can see that $\{(\overline{\mu}^*)^{p-1} \otimes 1\}(-v) = \iota_{2n} \otimes \iota_{2n} \otimes \ldots \otimes \iota_{2n}$ and as $((\overline{\mu}^*)^{p-1} \otimes 1)[\overline{H}^*(K(Z_p,2n)) \otimes \overline{H}^*(K(Z_p,2n))]$ in dim $2np$ is contained in

$[H^{2n}(K(Z_p,2n),Z_p)]^p$, $[(\bar{\mu}^*)^{p-1} \otimes 1](I_\beta) = 0$ in this dimension, $((\bar{\mu}^*)^{p-1} \otimes 1)z_\beta = 0$.

If $-v = \bar{\mu}^*d + z_\beta$

$$\iota_{2n} \otimes \ldots \otimes \iota_{2n} = ((\bar{\mu}^*)^{p-1} \otimes 1)(-v) = (\bar{\mu}^*)^p d.$$

Let $y \in H_{2n}(K(Z_p,2n),Z_p) = Z_p$ be a dual of ι_{2n} then, $\langle\iota_{2n},y\rangle = 1$. $1 = \langle\iota_{2n} \otimes \iota_{2n} \otimes \ldots \otimes \iota_{2n},\ y \otimes y \ldots \otimes y\rangle = \langle(\bar{\mu}^*)d,\ y \otimes \ldots \otimes y\rangle = \langle d,y^p\rangle$ contradicting the fact that $y^p = 0$.

To prove the first part suffices to show that for some $\hat{\xi}$ $HD(\hat{\xi}) = jv$. We shall prove this together with the second part.

One has the following pull back diagrams of ∞-loop spaces and maps.

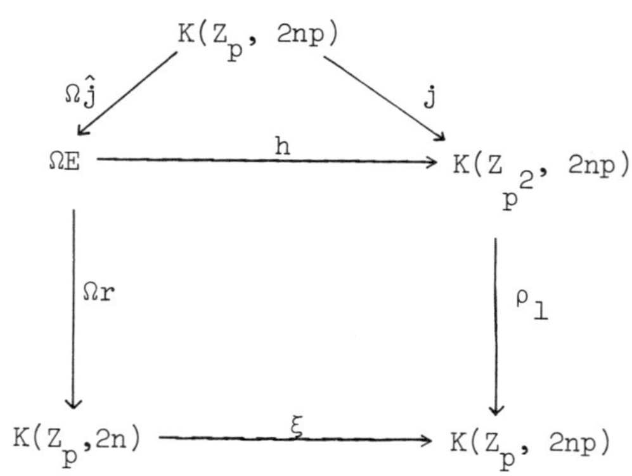

Choose arbitrary $\hat{\xi}$: $K(Z_p,2n) \longrightarrow K(Z_{p^2}, 2np)$. It yields $\hat{\chi}$: $K(Z_p,2n) \longrightarrow \Omega E$, $\Omega r\hat{\chi} = 1$, $h\hat{\chi} = \hat{\xi}$. $HD(\hat{\chi}) = \Omega\hat{j}v_1$ and $HD(\hat{\xi}) = \Omega\hat{j}v_1 = j[v + \bar{\mu}^*d + z_\beta]$. As $j[v + \bar{\mu}^*d + z_\beta - v_1] = 0$ $v + \bar{\mu}^*d + z_\beta - v_1$ factors through β: $K(Z_p, 2np-1) \longrightarrow K(Z_p, 2np)$, hence lies in I_β and one may assume $v_1 = v + \bar{\mu}^*d + z_\beta$.

$Add(\hat{\chi} \times \Omega\hat{j})$: $K(Z_p,2n) \times K(Z_p, 2np) \longrightarrow \Omega E$ is a homotopy equivalence and if μ_{v_1} is given by $\mu_{v_1}[(x,y), (\bar{x},\bar{y})] = x\cdot\bar{x},\ v_1(x,\bar{x})\cdot y\cdot\bar{y}$ then $\mathrm{Add}(\hat{X} \times \Omega\hat{j})$ is a μ_{v_1}, Add H-map hence an H-homotopy equivalence. As ΩE, Add is HA and HC so is μ_{v_1} and $H^*(K(Z_p,2n) \times K(Z_p,2np),Z_p)$, μ_{v_1} is co-commutative and co-associative but as the generator $1 \otimes \iota_{2np}$ cannot be represented by a primitive element it is not primitively generated in $\dim \leq 2np$. Consequently, if $0 \neq y \in H_{2n}(K(Z_p,2n),Z_p)$ $(y \otimes 1)^p \neq 0$ in $H_{2np}(K(Z_p,2n) \otimes K(Z_p,2np),Z_p)$ and there exists a class

u with $H^*(\mu_{v_1}, Z_p)u = u \otimes 1 + 1 \otimes u + \sum_{k=1}^{p-1} \frac{1}{p}\binom{p}{k}(\iota_{2n} \otimes 1)^k \otimes (\iota_{2n} \otimes 1)^{p-k}$,

$u = d_1 \otimes 1 + c(1 \otimes \iota_{2np})$, $c \neq 0$, $d_1 \in H^{2np}(K(Z_p,2n), Z_p)$ but in view of the coproduct of $1 \otimes \iota_{2np}$ $c = 1$, and one may assume $d = d_1$ $z_\beta = 0$. Altering $\hat{X}$ by d will then yield a lifting X $HD(X) = \Omega\hat{j}v$, $HD(\hat{\xi}) = jv$ and $(\Omega X \times j)$ is a μ_v-Add H homotopy equivalence.

Now one can prove 3.3.8 for $r = 1$ (p-odd).

<u>3.3.10. Proposition</u>: *Let* $x \in H^{2n}(X,Z_p)$, *p-odd. Then* x^p *and* $x^{p-1}\beta x$ *survive to* E_2 *in the Bockstein s.s. and* $\beta_2\{x^p\} = \{x^{p-1}\beta x\}$.

<u>Proof</u>: The fact that x^p and $x^{p-1}\beta x$ survive to E_2 is proved as in the case $r > 1$ (3.3.8). Again, it suffices to prove 3.3.10 for $X = K(Z_p,2n)$, $x = \iota_{2n}$. By 3.3.9 one has a lifting $\hat{\xi}: K(Z_p,2n) \longrightarrow K(Z_{p^2}, 2np)$ of $\xi: K(Z_p,2n) \longrightarrow K(Z_p,2np)$ so that

$HD(\hat{\xi}) = jv$, $v = \sum_{k=1}^{p-1} \frac{1}{p}\binom{p}{k}\ \iota_{2n}^k \otimes \iota_{2n}^{p-k}$. Let $\hat{\beta}_2: K(Z_{p^2},2np) \longrightarrow K(Z_p,2np+1)$ then $\hat{\beta}_2 j = \beta$, $j: K(Z_p,2np) \longrightarrow K(Z_{p^2},2np)$. One can see that $\beta v = \bar{\mu}^*(\iota_{2n}^{p-1} \cdot \beta\iota_{2n})$. It follows that $\hat{\beta}_2\hat{\xi}\iota_{2np+1} = \iota_{2n}^{p-1}\beta\iota_{2n} + x$,

$x \in PH^{2np+1}(K(Z_p,2n), Z_p)$. As $\{(\hat{\beta}_2\hat{\xi})^*\iota_{2np+1}\} = \{\beta_2\iota_{2n}^p\}$ the proof will be completed if one shows that $x \in \operatorname{im}\beta$. But on

$$PH^{odd}(K(Z_p, 2n), Z_p) \quad \operatorname{im}\beta = \ker\beta$$

As $\beta\hat{\beta}_2 = 0$, $\beta(\iota_{2n}^{p-1}\beta\iota_{2n}) = (p-1)\iota_{2n}^{p-2}(\beta\iota_{2n})^2 = 0$, $\beta x = 0$ and $x \in \operatorname{im}\beta$. 3.3.8 and 3.3.10 constitute Theorem 5.4 of [Browder]$_1$ for p odd.

The main result of [Browder]$_1$ is then given as follows:

3.3.11. Definition: Let A be a Hopf algebra over Z_p, $x \in A^{2n}$. A 1-*implication* of x is an element $x_1 \in A^{2np}$ so that either $x_1 = x^p$ or $\langle \bar{x}^p, x_1\rangle \neq 0$ for every $x \in A^*_{2n}$ (A^* the dual of A) with $\langle \bar{x}, x\rangle \neq 0$.

An m *implication* of $x \in A_{2n}$, $1 \leq m \leq \infty$ is a sequence $x = x_0, x_1, \ldots, x_m$ so that x_{i+1} is a 1-implication of x_i.

3.3.12. Browder's first implication theorem (6.1 in [Browder]$_1$):

Let X,μ *be an H-space,* E_r, β_r *the Bockstein s.s,* E^r *its dual homology Bockstein s.s. If* $x \in PE^r_{2n}$, $x = \beta^r y$ *then* x *has an* ∞ *implication.*

Outline of proof: One has to construct a sequence $x_0 = x, x_1, x_2, \ldots, x_m, \ldots \in E^r$ with x_i a 1-implication of x_{i-1}. The sequence x_i is constructed inductively as follows. One obtains a sequence $x_i \in PE^{r_i}_{2np^i}$, $r_0 = r$, $0 \leq r_i - r_{i-1} \leq 1$, x_i is represented in $E^{r_{i-1}}$ by a 1-implication of x_{i-1} and $x_i = \beta^{r_i} y_i$. One can pull back the x_i's to $E^{r_0} = E^r$ to obtain the desired implication sequence. By this inductive argument it suffices to construct x_1 out of $x = x_0$.

If $x^p = x_0^p \neq 0$ put $x_1 = x^p$, $r_1 = r_0$, $x^p = \beta_r(x^{p-1}y)$. So suppose $x^p = 0$ and suppose $<\bar{x},x> \neq 0$ $\bar{x} \in E_r^{2n}$. $\beta^r(x^{p-1}y) = x^p = 0$. One can prove now that $<\bar{x}^{p-1}\beta_r\bar{x}, x^{p-1}y> \neq 0$, hence, $\bar{x}^{p-1}\beta_r\bar{x} \notin \operatorname{im}\beta_r$ and by 3.2.8 and 3.2.10 $\beta_{r+1}\{\bar{x}^p\} = \{\bar{x}^{p-1}\cdot\beta_r\bar{x}\} \neq 0$ and $\{\bar{x}^p\} \neq 0$. $<\{\bar{x}^p\}, \beta^{r+1}\{x^{p-1}y\}> = <\bar{x}^{p-1}\beta_r\bar{x}, x^{p-1}y> \neq 0$, hence $\{x^{p-1}y\} \neq 0$ and computing the coproduct of $x^{p-1}y$ one can conclude that $\{x^{p-1}y\}$ and consequently $x_1 = \beta^{r+1}\{x^{p-1}y\} \neq 0$ are primitive. $<\{\bar{x}^p\}, x_1> \neq 0$ for $\bar{x}$, $<\bar{x},x> \neq 0$ implies that x_1 can be represented by a 1-implication in E^r. $x_1 = \beta^{r+1}y_1$, $y_1 = \{x^{p-1}y\}$.

This method yields further results, a sample of which we present here (for a more complete list and references, see [Browder]$_3$).

<u>3.3.13. Browder's implication theorems</u>: *Let* $\{E_n,\beta_n\}$ *be the* $H^*(\ ,Z)$ *Bockstein s.s and* X *an* H-*space.*

<u>3.3.13.1</u>. *If* $x \in P(E_r^{2m})$ $\beta_r x \neq 0$ *then* x *has an* ∞ *implication. Hence if* $H^*(X,Z_p)$ *is finite then* $\beta_r P(E_r^{2m}) = 0$.

<u>3.3.13.2.</u> *If* X *is* HC *and* $p = 2$ *and if* $x \in P(E_r^{2m})$ *then* x has an ∞ *implication. Hence, if* $H^*(X,Z_2)$ *is finite then* $PE_r^{2m} = 0$, *in particular* $PH^*(X,Z_2) \subset H^{odd}(X,Z_2)$ and $H^*(X,Z)$ *has no* 2-*torsions.*

<u>3.3.13.3</u>. *If* X *is* HC *and* HA *then* $x \in PE_r^{2m}$ *has an* ∞ *implication. Hence, if* $H^*(X,Z_p)$ *is finite* $H^*(X,Z)$ *has no* p-*torsion.*

3.4. High order operations.

The systematic study of high order operations started essentially in [Adams]$_1$. In general, let G, G' be abelian groups. *A (high order)* G-G' *cohomology operation* is a triple $\alpha = (x, E, y)$ where E is a space, $x \in H^*(E,G)$, $y \in H^*(E,G')$. Given a space K and a class $u \in H^*(K,G)$, u is in the domain of α if there exists $f\colon K \to E$ with $H^*(f,G)x = u$. The value $\alpha(u)$ is then defined as the set

$$\alpha(u) = \{H^*(f,G')y \in H^*(K,G') \mid f\colon K \to E,\ H^*(f,G)x = u\}.$$

One can easily replace x by a finite sequence $\overline{x} = x_1, \dots, x_m$, $x_i \in H^{n_i}(E,G)$, $\overline{y} = y_1, \dots, y_n$, $y_j \in H^{n_j}(E,G')$ and analogously define the domain of α and its values. In practice x, E, y are not chosen arbitrarily. Suppose $G = G' = Z_p$. One speaks of a $Z_p - Z_p$ *k-order operation* as (x,E,y) where E is a generalized k-stage Postnikov system:

$$\begin{array}{ccccc}
\Omega K_k & \xrightarrow{i_k} & E = E_k & & \\
 & & \downarrow{\scriptstyle r_k} & & \\
 & & E_{k-1} & \xrightarrow{f_k} & K_k \\
 & & \downarrow{\scriptstyle r_{k-1}} & & \\
 & & \vdots & & \\
 & & \downarrow & & \\
 & & E_1 = K_0 & \xrightarrow{f_1} & K_1
\end{array}$$

$\Omega K_j \xrightarrow{i_j} E_j \xrightarrow{r_j} E_{j-1}$ is the fibration induced by $f_j\colon E_{j-1} \to K_{j-1}$, $K_j = K(G_*)$ where G_* is a graded Z_p vector space.

x is then chosen to be the image of the vector of fundamental classes in $H^*(E_1,Z_p) = H^*(K_0,Z_p)$ and y is assumed to be a vector of classes not all in $\operatorname{im} H^*(r_k,Z_p)$.

If all spaces and maps are ∞ loop spaces and maps and y is an ∞-suspension, the operation is then called stable.

In this section we illustrate a technique used in the study of the cohomology of H-spaces involving a secondary operation (i.e.: an operation of order two). We shall give only the most elementary version. At the end of this section we shall list some of the stronger results obtained using this method, as well as results concerning the cohomology of H-spaces obtained by using operations in generalized cohomology theories.

We shall study only one secondary operation.

Let $K(Z_p,\ 2np-1) \xrightarrow{j_1} E_1 \xrightarrow{r_1} K(Z_p,2n)$ be the fibration induced by $\xi\colon K(Z_p,2n) \longrightarrow K(Z_p,2np)$ $H^*(\xi,Z_p)\iota_{2np} = \iota_{2n}^p$. Put $x = H^*(r_1,Z_p)\iota_{2n}$. Note that E_1 is an ∞ loop space.

<u>3.4.1. Lemma</u>: *There exists* $y \in H^{2np}(E_1,Z_p)$ *satisfying* $H^*(j_1,Z_p)y = \beta\iota_{2np-1}$ *and if* $x = H^*(r_1,r_p)\iota_{2n}$ *then*

$$(H^*(\mathrm{Add},\ Z_p) - p_1^* - p_2^*)y = \sum_{k=1}^{p-1} \frac{1}{p}\binom{p}{k}\ x^k \otimes x^{p-k}$$

<u>Proof</u>: Consider the stable diagram of induced fibrations:

$$\begin{array}{ccc}
\tilde{E}_1 & \longrightarrow & \tilde{E} \\
\downarrow{\scriptstyle \tilde{r}_1} & & \downarrow{\scriptstyle \tilde{r}} \\
K(Z_p,m) & = & K(Z_p,m) \\
\downarrow{\scriptstyle P^n} & & \downarrow{\scriptstyle \beta P^n} \\
K(Z_p,m+2n(p-1)) & \xrightarrow{\ \beta\ } & K(Z_p,m+2n(p-1)+1)
\end{array}$$

For $m = 2n$ one gets a diagram of ∞ loop spaces and maps

$$\begin{array}{ccc}
K(Z_p,2np-1) & \xrightarrow{\ \beta\ } & K(Z_p,2np) \\
\downarrow{\scriptstyle j_1} & & \downarrow \\
E_1 & \xrightarrow{\ g_1\ } & \Omega E \\
\downarrow{\scriptstyle r_1} & & \downarrow{\scriptstyle r} \\
K(Z_p,2n) & = & K(Z_p,2n)
\end{array}$$

By 3.3.9 $\Omega E \approx_H K(Z_p,2n) \times K(Z_p,2np)$ (the latter with a twisted multiplication μ_v).

In the notations of 3.3.9 one has an H-map

$g: E_1, \mathrm{Add} \longrightarrow K(Z_p,2n) \times K(Z_p,2np), \mu_v$ and $y = H^*(g,Z_p)(1 \otimes \iota_{2np})$ is the desired class.

Note that y is not uniquely determined by the properties of 3.4.1. It can be altered by any element in $H^*(r_1,Z_p)[PH^*(K(Z_p,2n), Z_p)]$.

Choose an arbitrary y satisfying the 3.4.1 properties and put $\alpha = (x, E_1, y)$. We would like to show that under certain conditions, if X, is an H-space, $u \in PH^{2n}(X,Z_p)$, $u^p = 0$ then $\alpha(u)$ consists of 1-implications of u (hence $0 \notin \alpha(u)$).

Given an H-space X,μ and $w: X \wedge X \longrightarrow K(Z_p,m)$ let μ_w be a multiplication on $X \times K(Z_p,m)$ given by

$$\mu_w[(x,y), (\overline{x},\overline{y})] = \mu(x,\overline{x}), \quad w(x,\overline{x})\cdot y \cdot \overline{y}$$

where $(\ )\cdot(\ )$ stands for the abelian group multiplication in $K(Z_p,m)$.

One can see that any multiplication on $X \times K(Z_p,m)$ so that both $p_1: X \times K(Z_p,m) \to X$ and $i_2: K(Z_p,m) \to X \times K(Z_p,m)$ are H-maps must be of this form. Indeed, if such $\widetilde{\mu}$ is given, $p_1\widetilde{\mu} = \mu p_1$ implies

$$\widetilde{\mu}[(x,y), (\overline{x},\overline{y})] = \mu(x,\overline{x}), \quad p_2\widetilde{\mu}[(x,y), (\overline{x},\overline{y})],$$

$$p_2\widetilde{\mu}: \quad X \times K(Z_p,m) \times X \times K(Z_p,m) \to K(Z_p,m).$$

For dimension reasons

$$H^*(p_2\widetilde{\mu}, Z_p)\iota_m = c_1(1 \otimes \iota_m \otimes 1 \otimes 1) + c_2(1 \otimes 1 \otimes 1 \otimes \iota_m) + \sum_i z_i' \otimes 1 \otimes z_i'' \otimes 1$$

As i_2 is an H-map, $c_1 = c_2 = 1$ and

$$p_2\widetilde{\mu}[(x,y), (\overline{x},\overline{y})] = \hat{w}(x,\overline{x})\cdot y \cdot \overline{y}; \quad \hat{w}: X \times X \to K(Z_p,m)$$

satisfies $H^*(w,Z_p)\iota_m = \sum_i z_i' \otimes z_i''$. Finally, $\widetilde{\mu}((x,*), (*,*)) = x,*$ and $\widetilde{\mu}((*,*), (x,*)) = x,*$, consequently $\hat{w}(x,*) = * = \hat{w}(*,x)$, $\hat{w} = w\Lambda$ and $\widetilde{\mu} = \mu_w$.

For any map $u: X \to K(Z_p,m)$ let $h_u: X \times K(Z_p,m) \to X \times K(Z_p,m)$ be given by $h_u(x,y) = x, u(x)\cdot y$. Then h_u are homeomorphisms $h_{u^{-1}} = h_u^{-1}$ and any self homotopy equivalence h of $X \times K(Z_p,m)$ with $p_1 h \sim p_1$ $hi_2 \sim i_2$ is homotopic to one of the h_u's.

We refer to such homotopy equivalences as $i_2 - p_1$ equivalences and it can be easily seen that if $\tilde{\mu}$ is an $i_2 - p_1 - \mu$ multiplication and h is an $i_2 - p_1$ equivalence with homotopy inverse g then $h\tilde{\mu}(g \times g)$ is an $i_2 - p_1 - \mu$ multiplication. Therefore, one can talk about essentially different $i_2 - p_1 - \mu$ multiplications. One has the following lemma:

<u>3.4.2. Lemma</u>: *The set of essentially different* $i_2 - p_1 - \mu$ *multiplications for* $X \times K(Z_p,m)$ *is in* 1-1 *correspondence with* $H^m(X \wedge X, Z_p)/\text{im}\,\bar{\mu}^*$.

<u>Proof</u>: Let $w: X \wedge X \longrightarrow K(Z_p,m)$, $u: X \to K(Z_p,m)$.

Consider

$$h_u \mu_w (h_{u^{-1}} \times h_{u^{-1}}): \; h_u \mu_w (h_{u^{-1}} \times h_{u^{-1}})[x,y), \bar{x},\bar{y})] =$$

$$= h_u \mu_w [(x, u(x)^{-1}\cdot y), \bar{x}, u(\bar{x})^{-1}\cdot\bar{y}] =$$

$$= h_u(\mu(x,\bar{x}), w(x,\bar{x})\cdot u(x)^{-1}\cdot u(\bar{x})^{-1}\cdot y\cdot\bar{y}] =$$

$$= \mu(x,\bar{x}), u(\mu(x,\bar{x}))u(x)^{-1}u(\bar{x})^{-1}w(x,\bar{x})y\cdot\bar{y}.$$

$$u(\mu(x,\bar{x}))\cdot u(x)^{-1}u(\bar{x})^{-1} = HD(u,\mu,m)[x,\bar{x}].$$

As $H^*(HD(u,\mu,m), Z_p) = \bar{\mu}^*(u)$ it follows that $h_u \mu_w (h_{u^{-1}} \times h_{u^{-1}}) = \mu_{\bar{\mu}^*(u)+w}$ and one has a surjection of $H^m(X \wedge X, Z_p)/\text{im}\,\bar{\mu}^*$ onto the set of essentially different $i_2 - p_1 - \mu$ multiplications. Conversely, if w_0, w_1 yield equivalent $i_2 - p_1 - \mu$ multiplications, $\mu_{w_1} = h_u \mu_{w_0}(h_{u^{-1}} \times h_{u^{-1}})$ for some $u: X \to K(Z_p, m)$ and $w_1 = \overline{\mu}^* u + w_0$.

3.4.3. Lemma: *Let* X, μ *be* HA. $w: X \wedge X \longrightarrow K(Z_p, m)$ *yields an HA structure* μ_w *on* $X \times K(Z_p, m)$ *if and only if* $(\overline{\mu}^* \otimes 1)w = (1 \otimes \overline{\mu}^*)w$ *where* $\overline{\mu}^*: \overline{H}^*(X, Z_p) \longrightarrow \overline{H}^*(X, Z_p) \otimes \overline{H}^*(X, Z_p)$ *is the reduced coproduct induced by* μ.

Proof:

$$\mu_w(\mu_w \times 1)[(x,y),(x',y')(x'',y'')] = \mu_w\{[\mu(x,x'),w(x,x')\cdot y\cdot y'],(x'',y'')\}$$

$$= (\mu(\mu \times 1))(x,x',x''),\ w(\mu(x,x'),x'') \cdot w(x,x')y\cdot y'\cdot y''$$

Similarly,

$$\mu_w(1 \times \mu_w)[(x,y),\ (x',y'),\ (x'',y'')] =$$

$$= (\mu(1 \times \mu))(x,x',x''),\ w(x,\ \mu(x',x'')) \cdot w(x',x'')\cdot y\cdot y\cdot y''.$$

$\mu_w(1 \times \mu_w) \sim \mu_w(\mu_w \times 1)$ if and only if the two maps

$$x,x',x'' \longrightarrow w(x,\ \mu(x',x'')) \cdot w(x',x'')$$

and

$$x,x',x'' \longrightarrow w(\mu(x,x'),\ x'') \cdot w(x,x')$$

are homotopic or equivalently if

$$x,x',x'' \longrightarrow w(x,\ \mu(x',x'')) \cdot w(x,x')^{-1} \cdot w(x,x'')^{-1}$$

$$x,x',x'' \longrightarrow w(\mu(x,x'),x'') \cdot w(x,x'')^{-1} \cdot w(x',x'')^{-1}$$

are homotopic. In functional notations:

$$w(1 \times \mu) - w(1 \times p_1) - w(1 \times p_2) \sim w(\mu \times 1) - w(p_1 \times 1) - w(p_2 \times 1)$$

and $(1 \otimes \overline{\mu}^*)w = (\overline{\mu}^* \otimes 1)w$.

3.4.4. Corollary: *The essentially different* $i_2 - p_1 - \mu$ *HA structures of* $X \times K(Z_p,m)$, *where* X,μ *is HA, are in* 1-1 *correspondence with*

$$\mathrm{Cotor}^{H^*(X,Z_p)}_{2,\,2m-2}(Z_p,Z_p)$$

Proof: Such H-structures are represented in view of 3.4.3 by elements of dim m in

$$\ker(\overline{\mu}^* \otimes 1 - 1 \otimes \overline{\mu}^*): \overline{H}^*(X,Z_p) \otimes \overline{H}^*(X,Z_p) \to$$
$$\to \overline{H}^*(X,Z_p) \otimes \overline{H}^*(X,Z_p) \otimes \overline{H}^*(X,Z_p)$$

and by 4.3.2 any two such structures are equivalent if and only if they differ by an element in $\mathrm{im}\ \overline{\mu}^*: \overline{H}^*(X,Z_p) \longrightarrow \overline{H}^*(X,Z_p) \otimes \overline{H}^*(X,Z_p)$. Hence, the essentially different $i_2 - p_1 - \mu$ HA structures on $X \times K(Z_p,m)$ are in 1-1 correspondence with $\ker(\overline{\mu}^* \otimes 1 - 1 \otimes \overline{\mu}^*)/\mathrm{im}\ \overline{\mu}^*$. But $\overline{\mu}^*$ and $\overline{\mu}^* \otimes 1 - 1 \otimes \overline{\mu}^*$ are the first two derivations in the co-bar construction determining $\mathrm{Cotor}^{H^*(X,Z_p)}(Z_p,Z_p)$.

3.4.5. Proposition: *Let* p *be an odd prime,* X,μ *an HA space with* $H^*(X,Z_p)$ *cocommutative. If* $x \in PH^{2m}(X,Z_p)$, $x^p = 0$ *then* $\alpha(x)$ *consists of* 1-*implications of* x.

Proof: Consider the diagram of H-spaces and H-maps:

$$
\begin{array}{ccccc}
 & & K(Z_p, 2mp-1) & & \\
 & & \downarrow j_1 & & \\
X \times K(Z_p, 2mp-1) & \xrightarrow{\mathrm{Add}(\chi \times j_1)} & E_1 & & \\
\downarrow p_1 & & \downarrow r_1 & & \\
X & \xrightarrow{x} & K(Z_p, 2m) & \xrightarrow{\xi} & K(Z_p, 2mp)
\end{array}
$$

$\chi\colon X \to E_1$ an arbitrary lifting of x.

Now ξ as an (∞) loop map is an HA map. By 2.5.2 $\theta(\xi x) = \Omega(\xi)\theta(x) + \theta(\xi)(x \wedge x \wedge x)$. As $\theta(\xi) = 0$ and $\Omega(\xi) \sim *$ $\theta(\xi x) = 0$ and $X \times K(Z_p, 2mp-1)$ with the (2.1.1) induced multiplication is an HA space. This multiplication being an $i_2 - p_1 - \mu$ multiplication is of the form μ_w, w classified by

$$\mathrm{Cotor}_{2,\ 2mp-3}^{H^*(X,Z_p)} (Z_p, Z_p)$$

As $H^*(X,Z_p)$ is co-associative $(X,\mu$ being HA) and cocommutative $PH^*(X,Z_p) \otimes PH^*(X,Z_p) \longrightarrow \mathrm{Cotor}_{2,*}^{H^*(X,Z_p)} (Z_p,Z_p)$ is an isomorphism in odd dimensions, hence, $w = \Sigma u_i' \otimes u_i'' + \bar{\mu}^* d$, $u_i', u_i'' \in PH^*(X,Z_p)$. It easily follows that $HD(\chi) = j_1 w$ and therefore

$$\bar{\mu}^* H^*(\chi,Z_p)y - [H^*(\chi,Z_p) \otimes H^*(\chi,Z_p)](\mathrm{Add})^* y = \beta w.$$

$$\bar{\mu}^* H^*(\chi,Z_p)y = \sum_{k=1}^{p-1} \frac{1}{p} \binom{p}{k} x^k \otimes x^{p-k} + \beta \Sigma u_i' \otimes u_i'' + \bar{\mu}^* \beta d.$$

Let $(\bar{\mu}^*)^2 = \bar{\mu}^*$ and $(\bar{\mu}^*)^m = [(\bar{\mu}^*)^{m-1} \otimes 1]\bar{\ }^*$ then the dual $(\bar{\mu}_*)^m$ of $(\bar{\mu}^*)^m$ satisfies $(\bar{\mu}_*)^m (z_1 \otimes z_2 \otimes \ldots \otimes z_m) = z_1 \cdot z_2 \cdot \ldots \cdot z_m$.

Let $z \in H_{2n}(X,Z_p)$ be any element. Then,

$$\langle H^*(\chi,Z_p)y, z^p\rangle = \langle(\bar{\mu}^*)^p H^*(\chi,Z_p)y, \underbrace{z \otimes z \otimes \ldots \otimes z}_{p}\rangle$$

As $u_i' \in PH^*(X,Z_p)$, $p > 2$ $(\bar{\mu}^*)^{p-1}u_i' = 0$ and $((\bar{\mu}^*)^{p-1} \otimes 1)(u_i' \times u_i'') = 0$.

One also has $\langle\beta d_1, z^p\rangle = \langle d_1, \beta_* z^p\rangle = 0$ for all d_1 and

$[(\bar{\mu}^*)^{p-1} \otimes 1] \sum_{k=1}^{p-1} \frac{1}{p}\binom{p}{k} x^k \otimes x^{p-k} = \underbrace{-x \otimes x \ldots \otimes x}_{p}$. Hence

$$\langle H^*(\chi,Z_p)y,z^p\rangle = \langle((\bar{\mu}^*)^{p-1} \otimes 1)\, \bar{\mu}^*H^*(\chi,Z_p)y, z \otimes z \otimes \ldots \otimes z\rangle$$

$$- \langle x \otimes x \ldots \otimes x,\ z \otimes z \otimes \ldots z\rangle + \langle(\bar{\mu}^*)^p\beta d\rangle$$

$$= -\langle x,z\rangle^p + \langle\beta d,\ z^p\rangle = -\langle x,z\rangle^p$$

3.4.5 now follows from the fact that $\alpha(x) = \{H^*(\chi,Z_p)y \mid \chi\colon X \to E_1, r_1\chi = x\}$.

<u>3.4.6. Example</u>: In [Harper]$_3$ an H-space X is constructed so that

$$H^*(X,Z_p) = \Lambda(x_3, x_{2p+1}) \otimes Z_p[x_{2p+2}]/(x_{2p+2}^p).$$

For dimension reasons x_i are primitive and there is no generator in dim $2p^2 + 2p$. 3.4.5 implies that X cannot be HA.

Another example of an application of this technique is the determination of the decomposability of the Steenrod operation P^p, p-odd.

<u>3.4.7. Proposition</u>: *There exist* $Z_p - Z_p$ *stable secondary operations* ϕ_1, ϕ_2 *so that*

$$P^{p-2}\phi_1 + \beta\phi_2 = P^p$$

(i.e., for any $x \in$ domain $\phi_1 \cap$ domain ϕ_2 $P^{p-2}[\phi_1(x)] + \beta[\phi_2(x)] = \{P^p x\}$.)

Proof: Consider the following matrices with entries in $A(p)$:

$$C = (P^{p-2}, \beta)$$

$$B = \begin{pmatrix} 2P^2 & -2P^1\beta + \beta P^1 \\ 0 & P^{p-1} \end{pmatrix} \qquad A = \begin{pmatrix} \beta \\ P^1 \end{pmatrix}$$

Then $CB = 0 = BA$. Let m be large and

$$f_A\colon K(Z_p,m) \longrightarrow K(Z_p, m+1) \times K(Z_p, m+2p-2)$$

$$f_B\colon K(Z_p, m+1) \times K(Z_p, m+2p-2) \to K(Z_p, m+4p-3) \times K(Z_p, m+2p(p-1))$$

$$f_C\colon K(Z_p, m+4p-3) \times K(Z_p, m+2p(p-1)) \to K(Z_p, m+2p(p-1)+1)$$

be given by:

$$H^*(f_A,Z_p)(\iota_{m+1} \otimes 1) = \beta\iota_m,\ H^*(f_A,Z_p)(1 \otimes \iota_{m+2p-2}) = P^1\iota_m$$

$$H^*(f_B,Z_p)(\iota_{m+4p-3} \otimes 1) = 2P^2\iota_{m+1} \otimes 1 + 1 \otimes (-2P^1\beta + \beta P^1)\iota_{m+2p-2}$$

$$H^*(f_B,Z_p)(1 \otimes \iota_{m+2p(p-1)}) = 1 \otimes P^{p-1}\iota_{m+2p-2}$$

$$H^*(f_C,Z_p)\iota_{m+2p(p-1)+1} = P^{p-2}\iota_{m+4p-3} \otimes 1 + 1 \otimes \beta\iota_{m+2p(p-1)}.$$

Then: $f_Bf_A \sim *$, $f_Cf_B \sim *$. Consider the following:

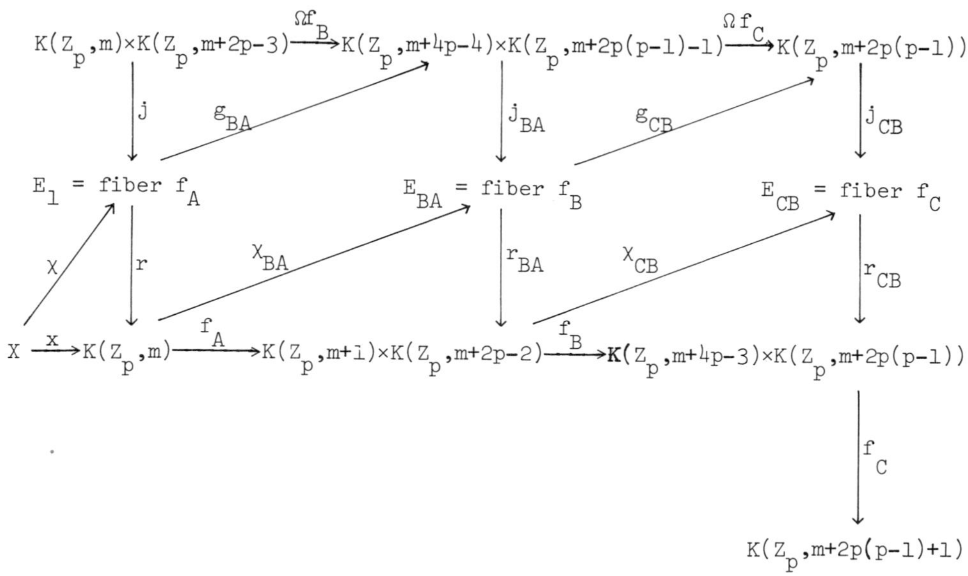

Define:

$$\phi_1 = (H^*(r,Z_p)\iota_m,\ E_1,\ H^*(g_{BA},\ Z_p)(\iota_{m+4p-4} \otimes 1)) = (x_1,E_1,y_1)$$

$$\phi_2 = (H^*(r,Z_p)\iota_m,\ E_1,\ H^*(g_{BA},\ Z_p)(1 \otimes \iota_{m+2p(p-1)-1})) = (x_1,E_1,y_2)$$

If $x \in H^m(X,Z_p)$ is in domain ϕ_1 $\cap$ domain ϕ_2 then

$$P^{p-2}\phi_1(x) + \beta\phi_2(x) = \{H^*(\Omega f_C g_{BA}\chi,\ Z_p)\iota_{m+2p(p-1)} | \chi\colon X \to E_1,\quad r\chi = x\}.$$

As two liftings χ_1 and χ of x will satisfy $\chi_1 = jv + \chi$ and as $\Omega f_C g_{BA} j = \Omega f_C \Omega f_B \sim *$ the set $P^{p-2}\phi_1(x) + \beta\phi_2(x)$ consists of the single element $H^*(g_{CB}\chi_{BA}x,\ Z_p)\iota_{m+2p(p-1)}$.

$H^*(g_{CB}\chi_{BA},\ Z_p)\iota_{m+2p(p-1)} \in H^{m+2p(p-1)}(K(Z_p,m),\ Z_p)$ which is Z_p-generated by $P^p\iota_m$, hence, $H^*(g_{CB}\chi_{BA},\ Z_p)\iota_{m+2p(p-1)} = cP^p\iota_m$, $c \in Z_p$. The proof will be completed if one shows that $c = 1$. For this it suffices to determine ϕ_i on any one class of any chosen space.

Fix an arbitrary n and let $\iota_{2n} \in H^{2n}(K(Z_p,2n), Z_p)$ be the fundamental class. Then $\iota_{2n}^p \in$ domain $\phi_1 \cap$ domain ϕ_2. We evaluate $\phi_i(\iota_{2n}^p)$.

Consider the diagram

$$\begin{array}{ccc}
E & \longrightarrow & E_1 \\
\downarrow{\scriptstyle \hat{r}} & & \downarrow{\scriptstyle r} \\
K(Z_p,\ 2n+1) & \xrightarrow{\ P^n\ } & K(Z_p,\ 2np+1) \\
\downarrow{\scriptstyle \beta P^n} & & \downarrow{\scriptstyle f_A} \\
K(Z_p,\ 2np+2) & \xrightarrow{\ i_1\ } & K(Z_p,\ 2np+2) \times K(Z_p,\ 2np + 2p-1)
\end{array}$$

As $P^1P^n = (n+1)P^{n+1}$, $P^{n+1}\iota_{2n+1} = 0$ and the diagram is commutative.

Looping the above diagram and applying 3.3.9 one obtains a diagram of H-spaces and maps:

$$\begin{array}{ccc}
K(Z_p,2np) & \xrightarrow{\ i_1\ } & K(Z_p,2np) \times K(Z_p,2np + 2p-3) \\
\downarrow{\scriptstyle i_2} & & \downarrow{\scriptstyle j_1} \\
K(Z_p,2n) \times K(Z_p,2np) & \xrightarrow{\ g\ } & E_1 \\
\downarrow & & \downarrow{\scriptstyle r_1} \\
K(Z_p,2n) & \xrightarrow{\ \xi\ } & K(Z_p,2np)
\end{array}$$

g is an H-map with respect to a multiplication μ_v on $K(Z_p,2n) \times K(Z_p,2np)$ and the loop multiplication on E_1.
$\mu_v[(x,y), (\overline{x},\overline{y})] = x\cdot\overline{x},\ v(x,\overline{x})\cdot y\cdot\overline{y}$, $v\colon K(Z_p,2n) \wedge K(Z_p,2n) \to K(Z_p,2np)$,

$v = \sum_{k=1}^{p-1} \frac{1}{p} \binom{p}{k} \iota_{2n}^k \otimes \iota_{2n}^{p-k}$. Put $\chi = gi_1\colon K(Z_p,2n) \to E_1$.

$HD(i_1, Add, \mu_v) = -i_2 v$, hence $HD(X, Add, Add) = -j_1 i_1 v$. Let μ = Add on $K(Z_p, 2n)$, μ_1 = Add on E_1. Then, as y_i are primitive

$$-\mu^* H^*(X, Z_p) y_1 = -\bar{\mu}_1^* H^*(g_{BA} X,\ Z_p)\iota_{2np+4p-4} \otimes 1 =$$

$$= H^*(g_{BA} j_1 i_1 v,\ Z_p)(\iota_{2np+4p-4} \otimes 1) =$$

$$= H^*(\Omega f_B i_1 v,\ Z_p)\iota_{2np+4p-4} \otimes 1 =$$

$$= 2P^2 \Sigma \frac{1}{p} \binom{p}{k} \iota_{2n}^k \otimes \iota_{2n}^{p-k}.$$

Similarly, $-\bar{\mu}^* H^*(X, Z_p) y_2 = H^*(\Omega f_B i_1 v,\ Z_p)(1 \otimes \iota_{2np+2p(p-1)-1}) = 0.$

Now one checks that

$$P^1 \sum_{k=1}^{p-1} \frac{1}{p} \binom{p}{k} \iota_{2n}^k \otimes \iota_{2n}^{p-k} = \bar{\mu}^* \iota_{2n}^{p-1} \cdot P^1 \iota_{2n}$$

and
$$2P^2 \sum_{k=1}^{p-1} \frac{1}{p} \binom{p}{k} \iota_{2n}^k \otimes \iota_{2n}^{p-k} = P^1 P^1 \sum_{k=1}^{p-1} \frac{1}{p} \binom{p}{k} \iota_{2n}^k \otimes \iota_{2n}^{p-k} =$$

$$= \bar{\mu}^*(-\iota_{2n}^{p-2}(P^1 \iota_{2n})^2 + 2\iota_{2n}^{p-1} P^2 \iota_{2n}) = \bar{\mu}^* d_0.$$

It follows that there exist primitives z_1, $z_2 \in H^*(K(Z_p, 2n), Z_p)$ so that $H^*(X, Z_p) y_1 = -d_0 + z_1 \in \phi_1(\iota_{2n}^p)$

$$H^*(X, Z_p) y_2 = z_2 \in \phi_2(\iota_{2n}^p)$$

To evaluate z_1 and z_2 consider

$$\sigma H^*(X, Z_p) y_1 = -\sigma d_0 + \sigma z_1 = \sigma z_1 \in PH^{2np+4p-5}(K(Z_p,\ 2n-1),\ Z_p)$$

$$\sigma H^*(X, Z_p) y_2 = \sigma z_2 \in PH^{2np+2p(p-1)-2}(K(Z_p,\ 2n-1),\ Z_p)$$

As $\Omega\xi \sim * \quad \Omega X = \Omega j_1 \hat{X}_1$.

$$\hat{X}_1\colon\ K(Z_p,\ 2n-1) \longrightarrow K(Z_p,\ 2np-1) \times K(Z_p,\ 2np+2p-4)$$

Put $u_1 = H^*(\hat{X}_1, Z_p)\iota_{2np-1} \otimes 1 \quad u_2 = H^*(\hat{X}_1, Z_p)(1 \otimes \iota_{2np+2p-4})$ then

$$\sigma z_1 = \Omega j_1[2P^2 u_1 + (-2P^1 + \beta P^1)u_2]$$

$$\sigma z_2 = \Omega j_1(P^{p-1}u_2)$$

As $H^*(K(Z_p,\ 2n-1),\ Z_p)$ is generated by $\sigma H^*(K(Z_p,2n),\ Z_p)$ $u_i = \sigma\hat{u}_i + d_i$ d_i - decomposable and as σz_i are indecomposable primitives it follows that $\sigma z_1 = \Omega j_1(2P^2\sigma\hat{u}_1 + (-2P^1\beta + \beta P^1)\sigma\hat{u}_2)$

$$\sigma z_2 = \Omega j_1(P^{p-1}\sigma\hat{u}_2)$$

and if one alters X by $j_1(\hat{u}_1 \times \hat{u}_2)\Delta_{K(Z_p,2n)}$ if necessary one may assume $\sigma z_i = 0$. As $[\ker \sigma\colon PH^*(K(Z_p,2n),\ Z_p) \to PH^*(K(Z_p,\ 2n-1),\ Z_p)] = 0$ in dim z_i one obtains $z_1 = z_2 = 0$, hence,

$$\iota_{2n}^{p-2} \cdot (P^1\iota_{2n})^2 - 2\iota_{2n}^{p-1}P^2\iota_{2n} \in \phi_1(\iota_{2n}^p) \qquad 0 \in \phi_2(\iota_{2n}^p).$$

$$P^{p-2}\phi_1(\iota_{2n}^p) + \beta\phi_2(\iota_{2n}^p) = P^{p-2}(\iota_{2n}^{p-2}(P^1\iota_{2n})^2 - 2\iota_{2n}^{p-1}P^2\iota_{2n}).$$

To evaluate the latter one uses the fact that it equals $P^p\iota_{2n}^p = c(P^1\iota_{2n})^p$ and to determine c one can make the calculations modulo the ideal generated by $P^i\iota_{2n}$, $i > 1$, and there

$$P^{p-2}(\iota_{2n}^{p-2}(P^1\iota_{2n})^2) = (P^1\iota_{2n})^p \text{ and } c = 1.$$

The decomposability of P^{p^r} $r > 1$ is more complex and proofs can be found in [Liulevicius].

Using the basic ideas of 3.4.5 a more refined but complex technique yields stronger results (see [Zabrodsky]$_{1,2,4,6}$ and [Lin]$_{1,2}$). Following is a sample of these results:

3.4.8. Proposition: *Let* X,μ *be an H-space.*

3.4.8.1. *If* $\beta H^*(X,Z_p) = 0$ p-*odd then there exists a homomorphism*

$$\phi: QH^{2m}(X,Z_p) \longrightarrow QH^{2mp}(X,Z_p)/\mathrm{im}\, P^1$$

with $\ker\phi \subset \mathrm{im}\, P^1$. ([Zabrodsky]$_6$)

3.4.8.2. *If* X,μ *is HC and* $H^*(X,Z_2)$ *is an exterior algebra on odd dimensional generators then* $Sq^{4n}: PH^{4n+1}(X,Z_2) \longrightarrow PH^{8n+1}(X,Z_2)$ *is a monomorphism.* ([Zabrodsky]$_4$)

In conjunction with the discoveries of piles of non-classical finite CW H-spaces the following was proved using similar methods:

3.4.9. Proposition ([Zabrodsky]$_5$): *Let* X *be an H-space,* $f: X \to S^n$ $n \neq 2^j-1$ *satisfy* $H^n(f,Z_2) \neq 0$. *Let* $h_\lambda: S^n \to S^n$ *be a map of degree* λ *and* $X(f,\lambda)$ *be the pullback of* f *and* h_λ. *If* λ *is even then* $X(f,\lambda)$ *does not admit an H-structure.*

Most recently, a long-standing conjecture was resolved for p-odd.

3.4.10. Proposition ([Lin)$_2$]: *Let* X,μ *be an H-space.* $H^*(X,Z_p)$ *finite,* p-*odd. Then* $H^*(\Omega X,Z)$ *is* p-*torsion free.*

Using K-theory operations, J. Hubbuck proved many cohomology structure theorems for H-spaces. (See [Hubbuck]$_{1-4}$) One of the most striking of his discoveries in this context is:

3.4.11. Proposition ([Hubbuck]$_2$): *Let* X *be finite dimensional HC space. Then* X *is homotopy equivalent to a torus* $T^n = (S^1)^n$.

Chapter IV

Mod p theory of H-spaces

Introduction

The source of any type of mod p theory in algebraic topology is Serre's study of classes of abelian groups, where an essential meaning was given to maps inducing isomorphisms modulo a given class of abelian groups (e.g.: the class of $\neq$ p-torsion groups).

The analysis of Serre's observations branches into two related but different methods: [Adams]$_2$ ignited the ideas leading to the localization theory formulated by Sullivan and developed to become a subfield in algebraic topology. (We shall use [Hilton, Mislin, Roitberg] as a standard reference to localization theory).

The other method in mod p theory is the one closer to the original work of Serre: [Mimura, Oneill, Toda] and [Mimura, Toda]$_{1,2}$ developed the theory of p-equivalences and p-universal spaces.

The main difference in the two theories is that the first carries us outside the category of (homotopy equivalents of) CW complexes of finite type which we try to avoid, on the other hand one has to concede that the mod p equivalence theory is effective only for a limited family of spaces (those called p-universal) while localization theory is universally effective.

Fortunately, H-spaces are p-universal for every p and for them mod p equivalence theory is equivalent to localization theory and it is the first one that we use here (the only exception is the proof of 4.2.5 where we refer to a localization theory proof in order to avoid the unnecessary development of its p-equivalence equivalent).

This chapter contains a survey of the basic facts in the [Mimura, Oneill, Toda] theory (section 1) as well as the notions of mod p homotopy due to Rector (section 2).

The decomposition of 0 equivalences of section 3 and the mixing technique of section 7 follow $[\text{Zabrodsky}]_{3,7}$ and lead to the study of the non classical H-spaces in section 8.

Mod p H-spaces are studied in section 5 with the main results being 4.5.2 and 4.5.3. The first was often assumed to be true though no explicit proof was produced as far as we know. 4.5.3 closes completely the uneasy gap in mixing H-spaces with nonconsistent rational Hopf algebras (and see remark 4.5.3.1).

In section 6 the notion of the genus of an H-space is studied and as an illustration the genus of SU(n) is given a lower bound.

4.1. p-equivalence and p-universal spaces

Let $\mathbb{P}$ denote the set of primes.

4.1.1. Definition: Let $p \in \mathbb{P} \cup \{0\}$. $f: X \to Y$ is said to be *a p-equivalence* if $H^*(f,Z_p)$, (equivalently $H_*(f,Z_p)$, $H^*(f,Z) \otimes Z_p$, $H_*(f,Z) \otimes Z_p$, $\pi(f) \otimes Z_p$, $\pi_*(f,Z_p)$) are isomorphisms. Here $Z_p = Z/pZ$ for $p \in \mathbb{P}$, $Z_0 = Q$.

f is said to be *a* $\mathbb{P}_1$ *equivalence*, $\mathbb{P}_1 \subset \mathbb{P}$, if f is a p-equivalence for every $p \in \mathbb{P}_1$.

Note that $f: X \to Y$ is a p-equivalence, $p \in \mathbb{P}$ if and only if $\pi_m(\text{fiber } f)$ are all torsion groups (of order prime to p) and hence every p equivalence is a 0-equivalence.

4.1.2. Definition: A space X is said to be $\mathbb{P}_1$ *Universal*, $\mathbb{P}_1 \subset \mathbb{P} \cup \{0\}$, if for every $q \notin \mathbb{P}_1$ and $n_0 > 0$ there exists $f = f_{q,n_0} : X \to X$ where f is a $\mathbb{P}_1$ equivalence and $H_n(f, Z_q) = 0$ for $0 < n \leq n_0$.

4.1.3. Definition: A space X is said to satisfy *a finiteness condition* if

(f.c.): (f.c.H) $H_n(X,Z) = 0$ for $n > N(X)$

or (f.c.π) $\pi_n(X) = 0$ for $n > N(X)$

The following only slightly differ from theorems 2.1 and 3.2 of [Mimura, Oneill, Toda].

4.1.4. Proposition: *For a space* X *the following are equivalent:*

(a) X *is* $\mathbb{P}_1$ *universal*

(b) For every $\mathbb{P}_1$*-equivalence* $f: X' \to X$ *with* fiber f *satisfying* (f.c.π) *there exists a* $\mathbb{P}_1$ *equivalence* $g: X \to X'$.

(c) There exists a $\mathbb{P}_1$ *equivalence* $f: X \to X_0$ *with* fiber f *satisfying* (f.c.π) *and* X_0 *is* $\mathbb{P}_1$ *universal.*

(d) For arbitrary spaces $Y, \tilde{Y}$, $\mathbb{P}_1$*-equivalence* $\tilde{f}: Y \to \tilde{Y}$ *with fiber satisfying* (f.c.π) *and a map* $\tilde{h}: X \to \tilde{Y}$, *there exist a* $\mathbb{P}_1$ *equivalence* $f: X \to X$ *and a map* $h: X \to Y$ *so that* $\tilde{h}f \sim \tilde{f}h$.

(e) For every integer n_0 *and a prime* $q \notin \mathbb{P}_1$ *there exists* $f = f_{n_0,q}: X \to X$ *so that* $\pi_n(f, Z_q) = 0$ *for* $0 < n \leq n_0$.

(f) *For every* $\mathbb{P}_1$ *equivalence* $f\colon X \to X'$ *with cofiber satisfying* (f.c.H) *there exists a* $\mathbb{P}_1$ *equivalence* $g\colon X' \to X$.

(g) *There exists a* $\mathbb{P}_1$ *equivalence* $f\colon X_0 \to X$ *with cofiber satisfying* f.c.H *and* X_0 *is* $\mathbb{P}_1$ *universal.*

(h) *For any* $\mathbb{P}_1$ *equivalence* $\tilde{f}\colon Y \to \tilde{Y}$ *with cofiber satisfying* (f.c.H) *and any map* $h\colon Y \to X$ *there exist a* $\mathbb{P}_1$ *equivalence* $f\colon X \to X$ *and a map* $\tilde{h}\colon \tilde{Y} \to X$ *so that* $\widetilde{h}\widetilde{f} \sim fh$.

<u>Proof</u>: The proof consists of the following steps:

$$(a) \Leftrightarrow (b), \quad (a) \Leftrightarrow (c), \quad (b) \Leftrightarrow (d), \quad (a) \Leftrightarrow (e),$$

$$(e) \Leftrightarrow (f), \quad (e) \Leftrightarrow (g) \quad \text{and} \quad (f) \Leftrightarrow (h).$$

(a) $\Rightarrow$ (b). Decompose $f\colon X' \to X$ into its primitive Postnikov decomposition (which is finite as fiber f satisfies (f.c.π) and involves only primes $q_i \notin \mathbb{P}_1$ as f is a $\mathbb{P}_1$ equivalence):

$$X' = X_m \xrightarrow{f_m} X_{m-1} \longrightarrow \cdots \xrightarrow{f_1} X_0 = X$$

Consider:

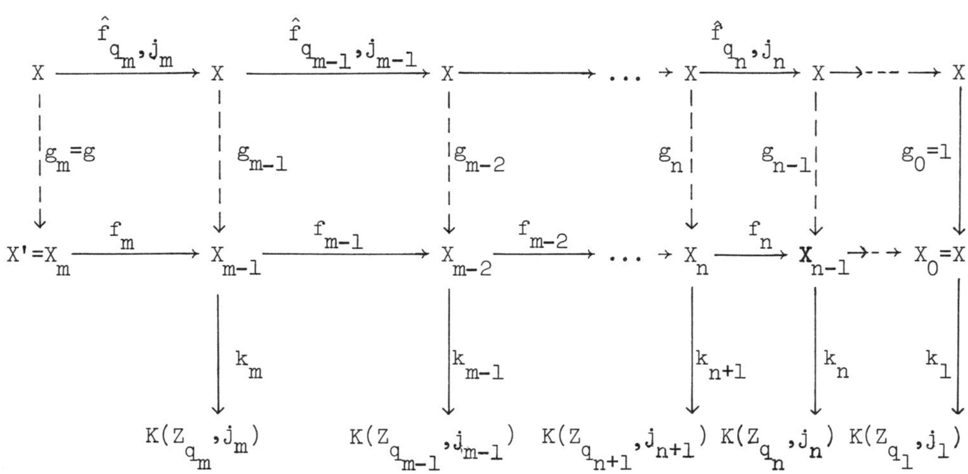

f_n - the fibration induced by k_n with fiber $K(Z_{q_n}, j_n-1)$,

$\hat{f}_{q_n,j_n} : X \to X$ is a $\mathbb{P}_1$ equivalence with $H^m(\hat{f}_{q_n,j_n}, Z_{q_n}) = 0$ for

$0 < m \leq j_n$. If g_i $i < n$ exist with $\hat{f}_{q_1,j_1}\hat{f}_{q_2,j_2}\ldots\hat{f}_{q_i,j_i} \sim f_1 f_2 \ldots f_i g_i$

then as $k_n g_{n-1}\hat{f}_{q_n,j_n} \sim *, g_n$ exists so that $f_n g_n \sim g_{n-1}\hat{f}_{q_n,j_n}$.

$\hat{f} = \hat{f}_{q_1,j_1}\hat{f}_{q_2,j_2}\ldots\hat{f}_{q_m,j_m}$ is a $\mathbb{P}_1$ equivalence, $fg_m \sim \hat{f}$ and $g_m = g$

is a $\mathbb{P}_1$ equivalence $g: X \to X'$.

(b) ⇒ (a). Given n_0 and $q \notin \mathbb{P}_1$ let $j: X \to \prod_{0<n\leq n_0} K(H^n(X,Z_q), n)$

yield a surjection of $H^m(\;, Z_q)$ $m \leq n_0$. Let $f: X' \to X$ be the fiber of

j. Then f is a $\mathbb{P}_1$ equivalence and fiber f satisfy (f.c.π). If

$g: X \xrightarrow{\approx \mathbb{P}_1} X'$ then $f_{q,n_0} = gf$ is a $\mathbb{P}_1$ equivalence and as

$H^m(f,Z_q) = 0$ for $0 < m \leq n_0$, $H^m(f_{q,n_0}, Z_p) = 0$ $0 < m \leq n_0$.

(a) ⇒ (c) trivial.

(c) ⇒ (a). Given n_0 and q let $\hat{f} = \hat{f}_{q,n_0} : X_0 \xrightarrow{\approx \mathbb{P}_1} X_0$ satisfy

$H^m(\hat{f}_{q,n_0}, Z_q) = 0$ $0 < m \leq n_0$. By (b) applied to X_0 there exists

$g: X_0 \xrightarrow{\approx \mathbb{P}_1} X$. Put $f_{q,n_0} = g\hat{f}f$.

(b) ⇒ (d). Given $\tilde{f}, \tilde{h}$ form their pull-back:

$$\begin{array}{ccc} \tilde{X} & \xrightarrow{h_1} & Y \\ \downarrow{\scriptstyle f_1} & & \downarrow{\scriptstyle \tilde{f}} \\ X & \xrightarrow{\tilde{h}} & \tilde{Y} \end{array}$$

As fiber $\tilde{f}$ = fiber f_1, f_1 is a $\mathbb{P}_1$ equivalence and fiber f_1 satisfies (f.c.π), by (b) there exists $g_1 \colon X \xrightarrow{\approx \mathbb{P}_1} \tilde{X}$. Put $f = f_1 g_1$, $h = h_1 g_1$.

(d) ⇒ (b). Given $f \colon X' \xrightarrow{\approx \mathbb{P}_1} X$ put $\tilde{Y} = X$, $Y = X'$, $\tilde{h} = 1$, $\tilde{f} = f$ and by (d) one has $f_1 \colon X \xrightarrow{\approx \mathbb{P}_1} X$ $1f_1 = fh$. As f, f_1 are $\mathbb{P}_1$ equivalences so is $h \colon X \to X'$.

(a) ⇒ (e) follows from the following more general observation:

If $f \colon X \to X$ satisfies $\tilde{H}^n(f, Z_q) = 0$ for $n \le n_0$ then there exists $m = m(n_0)$ so that $\pi_n(f^m, Z_q) = 0$ for $0 < n < n_0 - 2$.

Indeed, if $X = K(G,n)$, $H_m(f, Z_q) = 0$ $m \le n+1 \le n_0$ implies $H_m(f,Z) \otimes Z_q = 0$ and $\mathrm{Tor}(H_{m-1}(f,Z), Z_q) = 0$, therefore $\pi_n(f) \otimes Z_q = 0$ and $\mathrm{Tor}(\pi_n(f), Z_q) = 0$. But $\pi_n(K(G,n), Z_q) = \mathrm{Tor}(G, Z_q)$, $\pi_{n-1}(K(G,n), Z_q) = G \otimes Z_q$ and $\pi_m(K(G,n), Z_q) = 0$ for $m \ne n-1, n$, hence $\pi_*(f, Z_q) = 0$.

Suppose, inductively, for $n < n_0 - 1$ the existence of a positive integer $m = m(n)$ so that in the following commutative diagram

$$\begin{array}{ccc} X & \xrightarrow{\quad f^m \quad} & X \\ \downarrow{\scriptstyle h_n} & & \downarrow{\scriptstyle h_n} \\ Ht_n(X) = X_n & \xrightarrow{\quad (f_n)^m \quad} & X_n \end{array}$$

$H_j((f_n)^m, Z_q) = 0$ for $j \le n_0$ and $\pi_j((f_n)^m, Z_q) = 0$ for $j \le n+1 < n_0$.

Then as cofiber h_n is $n+1$ connected and H_{n+2} (cofiber h_n) $= \pi_{n+1}(X)$ one has

$$\begin{array}{ccccccc}
H_j(X,Z_q) & \xrightarrow{(h_n)_*} & H_j(X_n,Z_q) & \longrightarrow & H_j(\text{cofiber } h_n,Z_q) & \longrightarrow & H_{j-1}(X,Z_q) \\
\Big\downarrow f_*^m & & \Big\downarrow (f_n)_*^m = 0 & & \Big\downarrow (\tilde{f})_*^m & & \Big\downarrow f_*^m = 0 \\
H_j(X,Z_q) & \longrightarrow & H_j(X_n,Z_q) & \longrightarrow & H_j(\text{cofiber } h_n,Z_q) & \longrightarrow & H_{j-1}(X,Z_q)
\end{array}$$

and hence $(\tilde{f}^m)_j^2 = 0$, $j \leq n_0$ and in the following

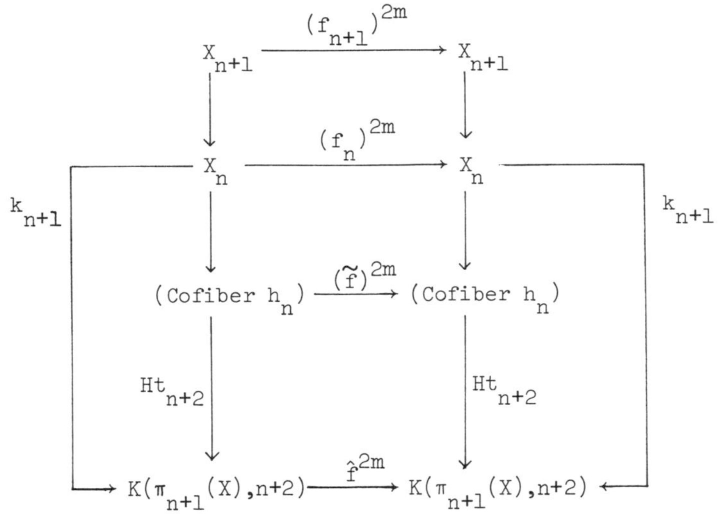

$$\begin{array}{ccc}
X_{n+1} & \xrightarrow{(f_{n+1})^{2m}} & X_{n+1} \\
\Big\downarrow & & \Big\downarrow \\
X_n & \xrightarrow{(f_n)^{2m}} & X_n \\
\Big\downarrow & & \Big\downarrow \\
(\text{Cofiber } h_n) & \xrightarrow{(\tilde{f})^{2m}} & (\text{Cofiber } h_n) \\
\Big\downarrow Ht_{n+2} & & \Big\downarrow Ht_{n+2} \\
K(\pi_{n+1}(X),n+2) & \xrightarrow{\hat{f}^{2m}} & K(\pi_{n+1}(X),n+2)
\end{array}$$

(with maps $k_{n+1}: X_n \to K(\pi_{n+1}(X),n+2)$ on both sides)

$H_j(\tilde{f}^{2m}, Z_q) = 0$, $j \leq n+2 \leq n_0$, imply $H_j(\hat{f}^{2m}, Z_q) = 0$, $j \leq n+2$ and $\pi_*(\hat{f}^{2m}, Z_q) = 0$. Together with $\pi_j((f_n)^{2m}, Z_q) = 0$, $j \leq n+1 < n_0$, this implies $\pi_j((f_{n+1})^{4m}, Zq) = 0$, $j \leq n+1 < n_0$. $H_{n+2}(\hat{f}^{2m},Z_q) = 0$ implies $H_*(\hat{f}^{2m}, Z_q) = 0$, ($H^*(K(G,k), Z_q)$ is generated as an algebra over the Steenrod algebra $A(q)$ by H^k). Now f_n^{4m}, f_{n+1}^{4m}, $\hat{f}^{4m}$, yield a comparison of the Serre spectral sequences of the fibrations

$$\begin{array}{ccc}
K(\pi_{n+1}(X), n+1) & \xrightarrow{(\Omega\hat{f})^{4m}} & K(\pi_{n+1}(X), n+1) \\
\downarrow & & \downarrow \\
X_{n+1} & \xrightarrow{(f_{n+1})^{4m}} & X_{n+1} \\
\downarrow & & \downarrow \\
X_n & \xrightarrow{(f_n)^{4m}} & X_n
\end{array}$$

$H_j((f_n)^{4m}, Z_q) = 0 = H_j((\Omega\hat{f})^{4m}, Z_q)$ $j \le n_0$ implies that the E^2 terms and consequently the E^∞ of the spectral sequences are mapped to each other by the 0-map. Consequently, in $\dim \le n_0$,

$0 = E_0[(f_{n+1})^{4m}]: E_0[H_*(X_{n+1}, Z_q)] \to E_0[H_*(X_{n+1}, Z_q)]$ where E_0 is the grading associated with the Serre filtration of $H_*(X_{n+1}, Z_q)$ and $H_j(f^{4m}_{n+1}, Z_q)$, $j \le n_0$, lowers filtration, hence, $H_j(f_{n+1}^{4m\cdot n_0}, Z_q) = 0$. Put $m(n+1) = 4m(n)\cdot n_0$ to complete the inductive step. Obviously

$$\pi_j(f^{m(n_0-2)}, Z_q) = \pi_j((f_{n_0-2})^{m(n_0-2)}, Z_q) = 0 \quad \text{for} \quad j < n_0-2.$$

One can dually prove that $\pi_j(f, Z_q) = 0$, $j \le n_0$ imply $H_j(f^m, Z_q) = 0$ $j < n_0-2$ to conclude (e) ⇒ (a). The proofs (e) ⇔ (f), (e) ⇔ (g) and (f) ⇔ (h) are the Eckmann-Hilton duals of (a) ⇔ (b), (a) ⇔ (c) and (b) ⇔ (d) respectively where pushouts replace pullbacks, Moore decompositions replace Postnikov approximation and cohomology and homotopy interchange.

<u>4.1.4. Corollary</u>: *If* X *is* $\mathbb{P}_1$ *universal so are* $Ht_n(X)$ *and* $H\ell_n(X)$. 4.1.3 *stays valid if the finiteness conditions imposed on* fiber f *in*

(b), (c) *and* (d) *and those for* cofiber f *in* (f), (g) *and* (h) *are replaced by the conditions that all spaces in* 4.1.3 *satisfy a finite condition of the same type* (f.c.π *or* f.c.H).

Proof: The $\mathbb{P}_1$ universality of $H\ell_n(X)$ follows directly from the definition of universal spaces 4.1.2. The $\mathbb{P}_1$ universality of $Ht_n(X)$ follows from 4.1.3 (d). The second part follows from the isomorphisms $[Ht_n(X), Ht_n(Y)] \xrightarrow{\approx} [H\ell_n(X), H\ell_{n+1}(Y)]$ and one can move from one to another in the proof of 4.1.3.

4.1.3 (d) easily implies that any H-space is $\mathbb{P}_1$ universal for all $\mathbb{P}_1$. But one has a stronger result:

4.1.6. Proposition ([Mimura, Toda]$_2$, Theorem 1.2): *Let* $h: X \to X_0$ *be a* 0-*equivalence. If* fiber h *satisfies* f.c.π *and* X_0 *is* $\mathbb{P}_1$ *universal so is* X.

Proof: Decompose h into a primitive decomposition.

$$X = X_m \xrightarrow{h_m} X_{m-1} \longrightarrow \cdots \xrightarrow{h_0} X_0$$

$h_j: X_j \to X_{j-1}$ induced by $k_j: X_{j-1} \to K(Z_{q_j}, n_j)$.

We shall use induction to show that X_j is $\mathbb{P}_1$ universal hence, it suffices to prove:

Let $h: X \to X_0$ be a $K(Z_p, n-1)$ principal fibration induced by $k: X \to K(Z_p, n)$. Then, if X_0 is $\mathbb{P}_1$ universal, so is X.
Fix $q \notin \mathbb{P}_1$ and n_0, $n_0 > n$. If $p \notin \mathbb{P}_1$ let $f_{n_0}: X_0 \to X_0$ be a $\mathbb{P}_1$ equivalence with $H_m(f_{n_0}, Z_{\hat{q}}) = 0$ for $\hat{q} = q$ and for $\hat{q} = p$, $m \leq n_0$.

Then one has a commutative diagram

$$\begin{array}{ccc}
X & \xrightarrow{\hat{f}} & X \\
\downarrow{\scriptstyle h} & & \downarrow{\scriptstyle h} \\
X_0 & \xrightarrow{f_{n_0}} & X_0 \\
\downarrow{\scriptstyle k} & & \downarrow{\scriptstyle k} \\
K(Z_p,n) & \xrightarrow{*} & K(Z_p,n)
\end{array}$$

As h is a $\mathbb{P}_1$ equivalence, f_{n_0} a $\mathbb{P}_1$ equivalence hence, $\hat{f}$ is a $\mathbb{P}_1$ equivalence. If $p \neq q$, $H^m(\hat{f},Z_q) = 0$ for $m \leq n_0$: $H^*(h,Z_q)H^*(f_{n_0},Z_q) = H^*(\hat{f},Z_q)H^*(h,Z_q)$, $H^*(h,Z_q)$ is an isomorphism and $H^m(f_{n_0},Z_q) = 0$ for $m \leq n_0$. If $p = q$ consider the Serre spectral sequence for homology in the diagram

$$\begin{array}{ccc}
K(Z_p,\ n-1) & \xrightarrow{*} & K(Z_p,\ n-1) \\
\downarrow{\scriptstyle j} & & \downarrow{\scriptstyle j} \\
X & \xrightarrow{\hat{f}} & X \\
\downarrow{\scriptstyle h} & & \downarrow{\scriptstyle h} \\
X_0 & \xrightarrow{f_{n_0}} & X_0
\end{array}$$

$E^2(\hat{f})$: $E^2(X) \sim H_*(X_0,Z_p) \otimes H_*(K(Z_p,\ n-1),\ Z_p) \to E^2(X)$

$E^2(\hat{f}) = H_*(f_{n_0},\ Z_p) \otimes H_*(*,Z_p)$ and $E^2_{m,n}(\hat{f}) = 0$, $0 < m+n \leq n_0$,

$E^\infty_{m,n}(\hat{f}) = 0$ and $\widetilde{H}_*(\hat{f},Z_p)$ lowers filtration in $\dim \leq n_0$. Hence, $\hat{f}^{n_0}$

is a $\mathbb{P}_1$ equivalence and $\widetilde{H}_m(\hat{f}^{n_0}, Z_p) = 0$ for $m \leq n_0$.

Suppose $p \in \mathbb{P}_1$. Let $f_{q,n_0} : X_0 \to X_0$ be a $\mathbb{P}_1$ equivalence, $H^n(f_{q,n_0}, Z_q) = 0$ for $0 < m \leq n_0$. $H^m(f_{q,n_0}, Z_p) \in GL(r,Z_p)$ $r = \text{rank}\ H^n(X_0,Z_p)$, hence there exists an integer a so that $H^n(f^a_{q,n_0}, Z_p) = 1$ and one has a commutative diagram

$$\begin{array}{ccc} X & \xrightarrow{\hat{f}} & X \\ \downarrow h & & \downarrow h \\ X_0 & \xrightarrow{f^a_{q,n_0}} & X_0 \\ \downarrow k & & \downarrow k \\ K(Z_p,n) & \xrightarrow{1} & K(Z_p,n) \end{array}$$

and as $H^*(h,Z_q)$ is an isomorphism, $H^*(f^a_{q,n_0}, Z_q) = 0 \Rightarrow H^*(\hat{f}, Z_q) = 0$. Finally, as $\hat{f}$ induces a homotopy equivalence fiber $h \to$ fiber h the upper square of the above diagram is a pull back diagram, fiber $\hat{f}$ = fiber f^a_{q,n_0} and $\hat{f}$ is a $\mathbb{P}_1$ equivalence.

4.1.7. Definition: A space X is called *an* H_0 *space if* $H^*(X,Q)$ is a free (associative, commutative graded) algebra. (e.g: Every H-space is an H_0-space).

4.1.8. Corollary: *An* H_0 *space* X *is* $\mathbb{P}_1$ *universal for every set* $\mathbb{P}_1$.

Proof: There exists a 0 equivalence $X \to K(\pi_*(X)/\text{torsion})$ and the latter is $\mathbb{P}_1$ universal, hence, 4.1.8 follows from 4.1.6.

4.2. Mod p homotopy

Most of the notions and properties discussed in this section are due to Rector (unpublished). First we formulate some simple properties of $\mathbb{P}_1$ equivalences some of which we have already used in the previous sections.

4.2.1. Lemma:

(a) Consider the following diagram:

4.2.1.1

$$\begin{array}{ccc} Y & \xrightarrow{\ h\ } & X \\ \downarrow{\scriptstyle g_1} & & \downarrow{\scriptstyle g_2} \\ Y_0 & \xrightarrow{\ h_0\ } & X_0 \end{array}$$

If 4.2.1.1 is either a pullback or a pushout diagram then h_0 *is a* $\mathbb{P}_1$ *equivalence if and only if* h *is.*

(b) If in 4.2.1.1 g_1, g_2 *are* $\mathbb{P}_1$ *equivalences and* h, h_0 *are* $\mathbb{P} - \mathbb{P}_1$ *equivalences then 4.2.1.1 is both a pullback and a pushout diagram.*

Proof: (a) A map f is a $\mathbb{P}_1$ equivalence if and only if $\pi_k(\text{fiber } f)$ are finite groups of order prime to $\mathbb{P}_1$ (or if and only if $H_k(\text{cofiber } f)$ are torsion groups of order prime to $\mathbb{P}_1$). If 4.2.1.1 is a pullback (pushout) diagram then h and h_0 have homotopy equivalent fibers (cofibers).

(b) Form the pullback of h_0, g_2 (pushout of h, g_1) and complete them to obtain:

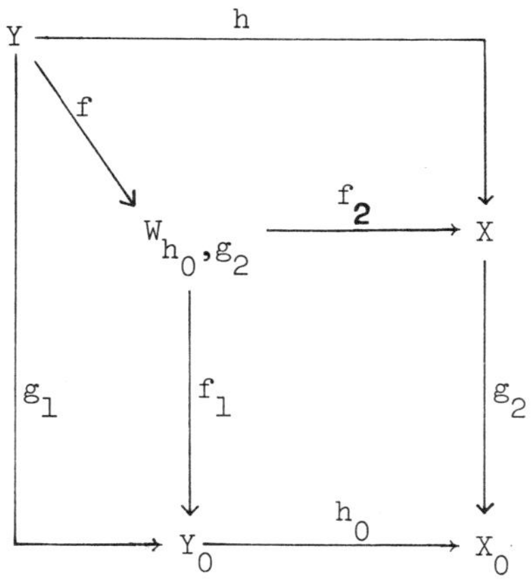

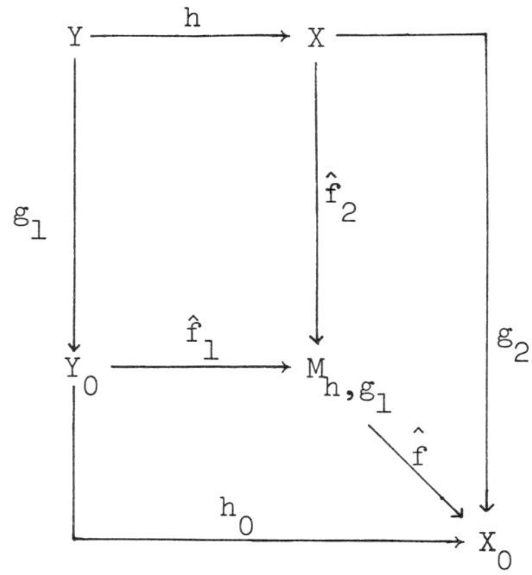

g_2 being $\mathbb{P}_1$ equivalence implies that so is f_1 and as g_1 is a $\mathbb{P}_1$ equivalence so is f. Similarly f is a $\mathbb{P} - \mathbb{P}_1$ equivalence. Being a p-equivalence for every p, f is a homotopy equivalence. Similarly for $\hat{f}$.

<u>4.2.2. Lemma (Rector)</u>: *Let* $f_1, f_2 \colon X \to Y$ *be maps. Then the following are equivalent:*

(a) There exists a $\mathbb{P}_1$ *equivalence* $g\colon X_1 \to X$ *with* $f_1 g \sim f_2 g$.

(b) There exists a $\mathbb{P}_1$ *equivalence* $g_1\colon Y \to Y_1$ *with* $g_1 f_1 \sim g_1 f_2$.

<u>Proof</u>: (a) ⇒ (b). For any maps $\alpha_1, \alpha_2\colon X \to Y$ let

$$X \overset{\alpha_1,\alpha_2}{\rightrightarrows} Y \xrightarrow{\ j\ } E_{\alpha_1,\alpha_2}$$

be an equalizer diagram (see e.g. [Spanier], p. 406). Consider the diagram

$$\begin{array}{ccccc} X_1 & \overset{f_1g,\ f_2g}{\rightrightarrows} & Y & \xrightarrow{j_1} & E_{f_1g,f_2g} \\ \downarrow g & & \| & & \downarrow \hat{g} \\ X & \overset{f_1,f_2}{\rightrightarrows} & Y & \xrightarrow{j} & E_{f_1,f_2} \end{array}$$

The cofiber of $\hat{g}$ is of the homotopy type of the suspension of the cofiber of g. Hence, $\hat{g}$ is a $\mathbb{P}_1$ equivalence. $f_1g \sim f_2g$ is equivalent to the existence of a left homotopy inverse for j_1, $h_1\colon E_{f_1g,f_2g} \to Y$. Consider

$$\begin{array}{ccccc} Y & \xrightarrow{j_1} & E_{f_1g,f_2g} & \xrightarrow{h_1} & Y \\ \| & & \downarrow \hat{g} & & \downarrow g_1 \\ Y & \xrightarrow{j} & E_{f_1,f_2} & \xrightarrow{h} & Y_1 \end{array}$$

with the right hand square being a pushout diagram. Then g_1 is a $\mathbb{P}_1$ equivalence and

$$g_1f_1 \sim g_1h_1j_1f_1 \sim hjf_1 \sim hjf_2 \sim g_1h_1j_1f_2 \sim g_1f_2$$

(b) $\Rightarrow$ (a) is dual (using co-equalizers).

<u>4.2.3. Definition (Rector)</u>: $f_1,f_2\colon X \to Y$ are said to be $\mathbb{P}_1$ *homotopic* ($f_1 \sim_{\mathbb{P}_1} f_2$) if one of the equivalent conditions of 4.2.2 are satisfied: $f_1g \sim f_2g$ for a $\mathbb{P}_1$ equivalence g or $g_1f_1 \sim g_1f_2$ for $\mathbb{P}_1$ equivalence g_1.

4.2.4. Lemma: $\sim_{\mathbb{P}_1}$ *is an equivalence relation respecting composition.*

Proof: $\sim_{\mathbb{P}_1}$ is obviously reflexive and symmetric.

Let $f_0, f_1, f_2 \colon X \to Y$, $f_0 \sim_{\mathbb{P}_1} f_1$, $f_1 \sim_{\mathbb{P}_1} f_2$. Let $g_1 \colon Y \to Y_1$, $g_2 \colon Y \to Y_2$ be $\mathbb{P}_1$ equivalences, $g_1 f_0 \sim g_1 f_1$, $g_2 f_1 \sim g_2 f_2$. Let $\hat{g}_1, \hat{g}_2, Y_0$ complete a pushout diagram for g_1, g_2

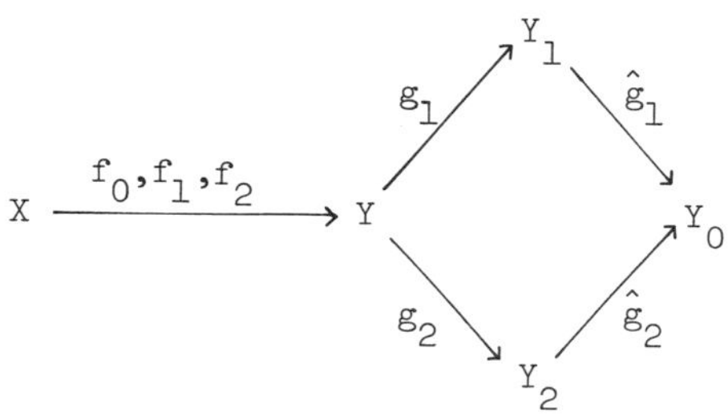

then $\hat{g}_i$ are $\mathbb{P}_1$ equivalences.

$$\hat{g}_2 g_2 f_0 \sim \hat{g}_1 g_1 f_0 \sim \hat{g}_1 g_1 f_1 \sim \hat{g}_2 g_2 f_1 \sim \hat{g}_2 g_2 f_2$$

and as $\hat{g}_2 g_2$ is a $\mathbb{P}_1$ equivalence $f_0 \sim_{\mathbb{P}_1} f_2$. Let $f_0, f_1 \colon X_0 \to X_1$, $f'_0, f'_1 \colon X_1 \to X_2$, $f_0 \sim_{\mathbb{P}_1} f_1$, $f'_0 \sim_{\mathbb{P}_1} f'_1$. Let $g \colon \hat{X}_0 \to X_0$ be a $\mathbb{P}_1$ equivalence $f_0 g \sim f_1 g$. Then $f'_0 f_0 g \sim f'_0 f_1 g$ and $f'_0 f_0 \sim_{\mathbb{P}_1} f'_0 f_1$. Let $g_1 \colon X_2 \xrightarrow{\approx \mathbb{P}_1} Y_2$, $g_1 f'_0 \sim g_1 f'_1$. Then $g_1 f'_0 f_1 \sim g_1 f'_1 f_1$ and $f'_0 f_1 \sim_{\mathbb{P}_1} f'_1 f_1$ and by transitivity $f'_0 f_0 \sim_{\mathbb{P}_1} f'_1 f_1$.

Using [Hilton, Mislin, Roitberg] Theorem II 5.3 and the fact that a $\mathbb{P}_1$ equivalence $\mathbb{P}_1$-localizes to a homotopy equivalence one has:

4.2.5. Proposition: *Given spaces* Y *and* X. *If* Y *satisfies* (f.c.H.) *or if* X *satisfies* (f.c.π) *and* $f_1, f_2 \colon Y \to X$ *then* $f_1 \sim_p f_2$ *for every* p *implies* $f_1 \sim f_2$.

4.3. Decomposition of 0 equivalences

We discuss in this section ideas introduced in [Zabrodsky]$_{3,7}$.

4.3.1. Proposition: *Let* $\psi\colon X \to X_0$ *be a* 0-*equivalence. Given a set of primes* $\mathbb{P}_1$, *there exists a space* $X(\mathbb{P}_1,\psi)$ *unique up to homotopy type and maps* $\psi'\colon X \to X(\mathbb{P}_1,\psi)$, $\psi''\colon X(\mathbb{P}_1,\psi) \to X_0$ *so that* $\psi''\psi' \sim \psi$, ψ' *is a* $\mathbb{P}_1$ *equivalence and* ψ'' *is a* $\mathbb{P} - \mathbb{P}_1$ *equivalence. Furthermore, given a commutative diagram*

$$\begin{array}{ccc} X & \xrightarrow{\quad\psi\quad} & X_0 \\ \Big\downarrow{\scriptstyle f} & & \Big\downarrow{\scriptstyle f_0} \\ \overline{X} & \xrightarrow{\quad\overline{\psi}\quad} & \overline{X}_0 \end{array}$$

$\psi,\overline{\psi}$ 0-*equivalences,* $\mathbb{P}_1$ - *a set of primes, then there exists a map* $f(\mathbb{P}_1,\psi,\overline{\psi})$, *so that the following is commutative:*

$$\begin{array}{ccccc} X & \xrightarrow[\approx\,\mathbb{P}_1]{\quad\psi'\quad} & X(\mathbb{P}_1,\psi) & \xrightarrow[\approx\,\mathbb{P}-\mathbb{P}_1]{\quad\psi''\quad} & X_0 \\ \Big\downarrow{\scriptstyle f} & & \Big\downarrow{\scriptstyle f(\mathbb{P}_1,\psi,\overline{\psi})} & & \Big\downarrow{\scriptstyle f_0} \\ \overline{X} & \xrightarrow[\approx\,\mathbb{P}_1]{\quad\overline{\psi}'\quad} & \overline{X}(\mathbb{P}_1,\overline{\psi}) & \xrightarrow[\approx\,\mathbb{P}-\mathbb{P}_1]{\quad\overline{\psi}''\quad} & \overline{X}_0 \end{array}$$

If both X *and* X_0 *satisfy* (f,c.H) *or* $\overline{X},\overline{X}_0$ *satisfy* (f.c.π) *then* $f(\mathbb{P}_1,\psi,\overline{\psi})$ *is unique up to homotopy.*

Proof: $X(\mathbb{P}_1,\psi)$, ψ' and ψ'' are constructed inductively as follows:

Suppose X_n, $\psi_n'\colon X \to X_n$, $\psi_n''\colon X_n \to X_0$ are given so that, ψ_n',ψ_n'' are 0-equivalences, $\psi_n''\psi_n' \sim \psi$, ψ_n'' is a $\mathbb{P}$-$\mathbb{P}_1$ equivalence and for

$p \in \mathbb{P}_1$ $H_m(\psi_n', Z_p)$ is an isomorphism for $m < n$ and an epimorphism for $m = n$, or equivalently $H_m(\text{cofiber } \psi_n', Z)$ are torsion groups of order prime to $\mathbb{P}_1$ for $m \leq n$.

Let $V_n = \text{cofiber } \psi_n'$. Consider the compositions k_{n+1} $X_n \to V_n \to V_n/H\ell_n(V_n) \to K(H_{n+1}(V_n), n+1) \to K({}^{\mathbb{P}_1}H_{n+1}(V_n), n+1)$ where ${}^{\mathbb{P}_1}H_{n+1}(V_n)$ is the sum of the p-primary components of

$$H_{n+1}(V_n, Z) = H_{n+1}(V_n), \quad p \in \mathbb{P}_1.$$

Note that as ψ_n' is a 0-equivalence $H_{n+1}(V_n, Z)$ is torsion and k_{n+1} is natural. Obviously $k_{n+1}\psi_n' \sim *$ and if $X_{n+1} = \text{fiber } k_{n+1}$ ψ_n' lifts to $\psi_{n+1}': X \to X_{n+1}$. Put $\psi_{n+1}'' = \psi_n'' h_{n+1,n}$, $h_{n+1,n}: X_{n+1} \to X_n$. $h_{n+1,n}$ is obviously a $\mathbb{P}$-$\mathbb{P}_1$ equivalence, hence so is ψ_{n+1}''.

$$\psi_{n+1}''\psi_{n+1}' = \psi_n'' h_{n+1,n}\psi_{n+1}' = \psi_n''\psi_n' = \psi.$$

Finally, $H_m(\text{cofiber } \psi_{n+1}', Z) = H_m(\text{cofiber } \psi_n', Z)$ for $m \leq n$ and $H_{n+1}(\text{cofiber } \psi_{n+1}', Z) = {}^{\mathbb{P}-\mathbb{P}_1}H_{n+1}(\text{cofiber } \psi_n', Z)$ which completes the inductive step. $X(\mathbb{P}_1, \psi) = \underleftarrow{\lim}\, X_n$, $\psi' = \underleftarrow{\lim}\, \psi_n'$ and $\psi'': X(\mathbb{P}_1, \psi) \to X_0$ - the composition $X(\mathbb{P}_1, \psi) \xrightarrow{\alpha_n} X_n \xrightarrow{\psi_n''} X_0$ (which is independent of n) are the desired space and maps. The uniqueness of homotopy type will follow from the second part of the proposition.

The existence of $f(\mathbb{P}_1, \psi, \overline{\psi})$ follows from the following

<u>4.3.1.1. Lemma</u>: *Given a commutative diagram*

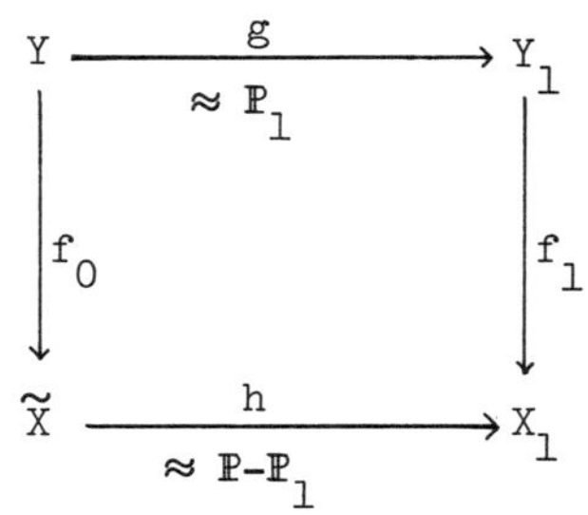

There exists $\tilde{f}: Y_1 \to \tilde{X}$ *so that* $\tilde{f}g \sim f_0$, $h\tilde{f} \sim f_1$.

Proof of 4.3.1.1: Construct the pull back $W_{f_1,h}$ of f_1 and h and complete:

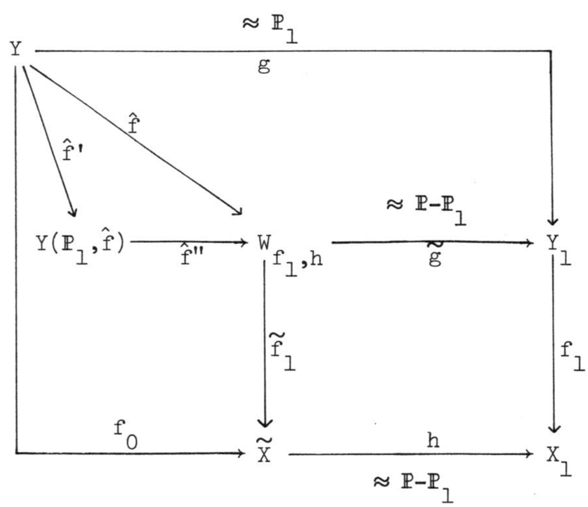

As $\tilde{g}$ is a $\mathbb{P}$-$\mathbb{P}_1$ equivalence (by 4.2.1 (a)) and g is a $\mathbb{P}_1$ equivalence $\hat{f}$ is a 0-equivalence and by the first part of 4.3.1 $Y(\mathbb{P}_1,\hat{f})$, $\hat{f}'$ and $\hat{f}''$ exist. $\hat{f}''$ and $\tilde{g}$ are $\mathbb{P}$-$\mathbb{P}_1$ equivalence and so is $\widetilde{g\hat{f}''}$. $\hat{f}'$ and g are $\mathbb{P}_1$-equivalences, hence, so is $\widetilde{g\hat{f}''}$ and $\widetilde{g\hat{f}''}$ is a homotopy equivalence. One has

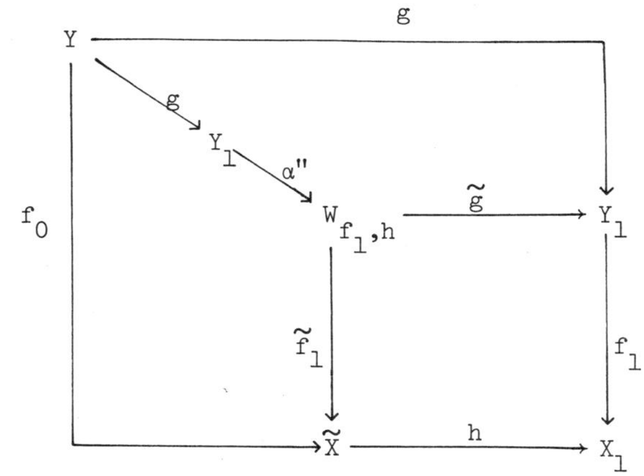

$\tilde{g}\alpha'' \sim 1$. (Simply choose $\alpha'' = \hat{f}''u$ where u is the homotopy inverse of $\widetilde{\hat{g}f''}$). $\tilde{f} = \tilde{f}_1\alpha''$ is the desired map: $\tilde{f}g = \tilde{f}_1\alpha''g \sim f_0$, $h\tilde{f} = h\tilde{f}_1\alpha'' \sim f_1\tilde{g}\alpha'' \sim f_1$.

Now, apply 4.3.1.1 for $Y = X$, $Y_1 = X(\mathbb{P}_1,\psi)$, $\tilde{X} = \bar{X}(\mathbb{P}_1,\bar{\psi})$, $X_1 = \bar{X}_0$, $g = \psi'$, $h = \bar{\psi}''$, $f_0 = \bar{\psi}'f$, $f_1 = f_0\psi''$ to obtain $f(\mathbb{P}_1,\psi,\bar{\psi})$. The uniqueness of homotopy type of $X(\mathbb{P}_1,\psi)$ follows from 4.3.1.1:

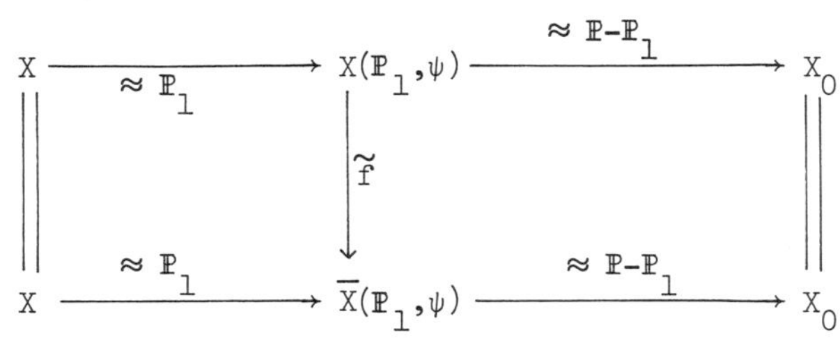

and $\tilde{f}$ is a $\mathbb{P}_1$ and a $\mathbb{P}-\mathbb{P}_1$ equivalence, hence a homotopy equivalence. Finally, $f(\mathbb{P}_1,\psi,\bar{\psi})$ is unique in the f.c. case by 4.2.5: If given

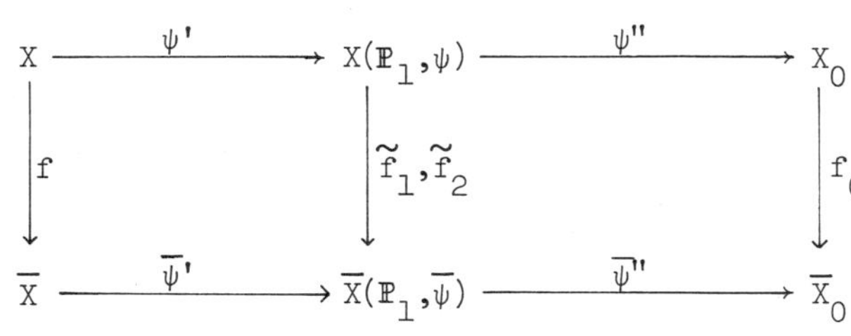

$\overline{\psi}''\tilde{f}_1 \sim \overline{\psi}''\tilde{f}_2$ implies $\tilde{f}_1 \sim_{\mathbb{P}-\mathbb{P}_1} \tilde{f}_2$ and $\tilde{f}_1\psi' \sim \tilde{f}_2\psi'$ implies $\tilde{f}_1 \sim_{\mathbb{P}_1} \tilde{f}_2$ and by 4.2.5 $\tilde{f}_1 \sim \tilde{f}_2$, (note that by uniqueness of the homotopy type of $X(\mathbb{P}_1,\psi)$ and $\overline{X}(\mathbb{P}_1,\overline{\psi})$ they inherit the f.c. of X,X_0 and, $\overline{X},\overline{X}_0$ respectively).

<u>4.3.2. Corollary</u>: *If* X,X_0 *are* H-*spaces,* $\psi: X \xrightarrow{\approx 0} X_0$ *is an* H-*map then* $X(\mathbb{P}_1,\psi)$, ψ' *and* ψ'' *admit* H-*structures.*

<u>Proof</u>: By the uniqueness of the 4.3.1 decomposition one has $(X \times \overline{X})(\mathbb{P}_1,\psi \times \overline{\psi}) \approx X(\mathbb{P}_1,\psi) \times \overline{X}(\mathbb{P}_1,\overline{\psi})$ and if X,μ, X_0,μ_0 are H-spaces, ψ an H-map and a 0-equivalence one has a commutative diagram

$$\begin{array}{ccccc}
X \times X & \xrightarrow{\psi' \times \psi'} & X(\mathbb{P}_1,\psi) \times X(\mathbb{P}_1,\psi) & \xrightarrow{\psi'' \times \psi''} & X_0 \times X_0 \\
\downarrow{\scriptstyle \mu} & & \downarrow{\scriptstyle \mu(\mathbb{P}_1,\psi \times \psi,\psi)} & & \downarrow{\scriptstyle \mu_0} \\
X & \xrightarrow{\psi'} & X(\mathbb{P}_1,\psi) & \xrightarrow{\psi''} & X_0
\end{array}$$

Let $i_\varepsilon: X(\mathbb{P}_1,\psi) \to X(\mathbb{P},\psi) \times X(\mathbb{P}_1,\psi)$ $\varepsilon = 1,2$ be the injections. Then by 4.3.1 $\mu(\mathbb{P}_1, \psi \times \psi, \psi)i_\varepsilon = \alpha_\varepsilon$ are homotopy equivalences

$\varepsilon = 1,2$, $\alpha_\varepsilon\psi' \sim \psi'$, $\psi''\alpha_\varepsilon \sim \psi''$. Let γ_ε be the homotopy inverses of α_ε, $\gamma_\varepsilon\psi' \sim \psi'$ and $\psi''\gamma_\varepsilon \sim \psi''$. Then one can replace $\mu(\mathbb{P}_1,\psi \times \psi,\psi)$ by $\mu(\mathbb{P}_1,\psi \times \psi,\psi)(\gamma_1 \times \gamma_2)$, which is an H-structure for $X(\mathbb{P}_1,\psi)$ and $\mu(\mathbb{P}_1,\psi \times \psi,\psi)(\gamma_1 \times \gamma_2)(\psi' \times \psi') \sim \mu(\mathbb{P}_1,\psi \times \psi,\psi)(\psi' \times \psi') \sim \psi'\mu$. $\psi''\mu(\mathbb{P}_1,\psi \times \psi,\psi)(\gamma_1 \times \gamma_2) \sim \mu_0(\psi'' \times \psi'')(\gamma_1 \times \gamma_2) \sim \mu_0(\psi'' \times \psi'')$ and both ψ' and ψ'' are H-maps.

<u>4.3.3. Lemma</u>: *Let* $\psi: X \xrightarrow{\approx 0} X_0$ *and let* $\{\mathbb{P}_i \mid i = 1,\ldots,k\}$ *be a partition of* $\mathbb{P}$. *Then* X *is the pullback of the maps* $\psi_i'': X(\mathbb{P}_i,\psi) \to X_0$.

<u>Proof</u>: We prove that the pullback of $\psi_i'': X(\mathbb{P}_i,\psi) \to X_0$ and $\psi_j'': X(\mathbb{P}_j,\psi) \to X_0$ is $X(\mathbb{P}_i \cup \mathbb{P}_j,\ \psi)$. Indeed, consider the pullback of ψ_i'' and ψ_j'' and complete

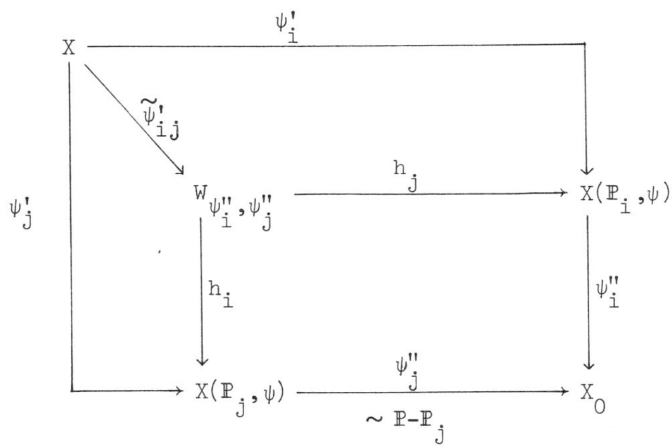

As ψ_j'' is a $\mathbb{P}$-$\mathbb{P}_j$ equivalent so is h_j. Similarly h_i is a $\mathbb{P}$-$\mathbb{P}_i$ equivalence. $\tilde{\psi}_{ij}'' = \psi_i'' h_j$ is a $(\mathbb{P}-\mathbb{P}_i) \cap (\mathbb{P}-\mathbb{P}_j)$ equivalence while as

$\mathbb{P}-\mathbb{P}_j \supset \mathbb{P}_i$ h_j is a $\mathbb{P}_i$ equivalence and so is ψ_i'. Consequently $\tilde{\psi}_{ij}'$ is a $\mathbb{P}_i$ equivalence. Similarly $\tilde{\psi}_{ij}'$ is a $\mathbb{P}_j$ equivalence, one gets

$$X \xrightarrow[\approx \mathbb{P}_i \cup \mathbb{P}_j]{\psi_{ij}} W_{\psi_i'',\psi_j''} \xrightarrow[\approx \mathbb{P} - (\mathbb{P}_i \cup \mathbb{P}_j)]{\tilde{\psi}_{ij}''} X_0$$

and by 4.3.1

$$W_{\psi_i'',\psi_j''} \approx X(\mathbb{P}_i \cup \mathbb{P}_j,\ \psi).$$

4.4. A Study of H_0-Spaces.

Recall that an H_0 space is a space X with a free rational cohomology algebra. If $G_n \approx QH^n(X,Z)/\text{torsion}$ and $\alpha_n: G_n \to H^n(X,Z)$ is a left inverse of the projection $H^n(X,Z) \to G_n$ then α_n can be realized by $\psi_n: X \to K(G_n,n)$. If X is an H_0 space then $\psi: X \to K(G_*) = \prod_n K(G_n,n)$ induced by ψ_n is a 0-equivalence and induces an isomorphism $QH^*(\psi,Z)/\text{torsion}$. Note that for an H_0 space X $QH^*(X,Z)/\text{torsion} \approx \pi_*(X)/\text{torsion}$. Hence, H_0 spaces have rational models in the form of $K(G_*) = \prod_n K(G_n,n)$ where $G_* = \{G_n\}$ is a free graded abelian group.

Here are some simple observations and notations regarding $K(G_*)$ and H_0-spaces.

There exists a natural monomorphism

$$\Gamma: \text{Hom}^0(G_*,G'_*) \to [K(G_*), K(G'_*)].$$

Furthermore, if $\phi(G_*)$ is the free (associative commutative graded) ring generated by G_* there exists an injection

$$\hat{\Gamma}: \hat{\text{Hom}}(\phi(G_*), \phi(G'_*)) \to [K(G_*), K(G'_*)]$$

where $\hat{\text{Hom}}$ stands for the set of ring homomorphisms (one has an obvious injection $\phi: \text{Hom}(G_*,G'_*) \to \hat{\text{Hom}}(\phi(G_*), \phi(G'_*))$.

For any H_0 space X and $\chi: QH^*(X,Z)/\text{torsion} \to H^*(X,Z)$ so that $QH^*(X,Z)/\text{torsion} \xrightarrow{\chi} H^*(X,Z) \xrightarrow{\text{proj}} QH^*(X,Z)/\text{torsion}$ is a monomorphism of maximal rank there exists $\psi = \psi(\chi): X \xrightarrow{\approx 0} K(QH^*(X,Z)/\text{torsion})$ realizing χ. We shall call such a χ *a rational splitting*. Such χ induces a morphism

$\hat{\chi}$: $\phi(QH^*(X,Z)/\text{torsion}) \to H^*(X,Z)$. Let $f: X \to X'$ be a map between H_0 spaces and given rational splittings:

$$\chi: QH^*(X,Z)/\text{torsion} \to H^*(X,Z)$$

$$\chi: QH^*(X',Z)/\text{torsion} \to H^*(X',Z)$$

For a vector $\overline{\lambda} = (\lambda_1, \lambda_2, \ldots)$ of integers let $I_{\overline{\lambda}} \in \mathrm{Hom}^0(G_*, G_*)$ be the endomorphism $(I_{\overline{\lambda}})_n = \lambda_n \cdot 1: G_n \to G_n$. Then one can always choose a vector $\overline{\lambda} = (\lambda_1, \ldots, \lambda_n, \ldots)$ of integers and a morphism

$$\alpha_0: \phi(QH^*(X',Z)/\text{torsion}) \to \phi(QH^*(X,Z)/\text{torsion})$$

so that $H^*(f,Z)\phi(X')\phi(I_{\overline{\lambda}}) = \phi(X)\phi(I_{\overline{\lambda}})\alpha_0$.

The geometric realization of this observation is given by:

4.4.2. Lemma:

(a) *Let* X *be an* H_0*-space. The rational splittings* $QH^*(X,Z)/\text{torsion} \to H^*(X,Z)$ *are in* 1-1 *correspondence with the rational equivalences* $\psi: X \to X_0 = K(QH^*(X,Z)/\text{torsion})$.

(b) *For any map* $f: X \to X'$ *and rational equivalences* $\psi: X \to X_0 = K(QH^*(X,Z)/\text{torsion})$, $\psi': X' \to X_0' = K(QH^*(X',Z)/\text{torsion})$ *there exists a vector* $\overline{\lambda} = (\lambda_1, \lambda_2, \ldots, \lambda_n, \ldots)$ *of integers and a map* $f_0: X_0 \to X_0'$ $(QH^*(f_0,Z)/\text{torsion} = QH^*(f,Z)/\text{torsion})$ *so that*

$$\Gamma(I_{\overline{\lambda}})\psi' f \sim f_0 \Gamma(I_{\overline{\lambda}})\psi.$$

4.4.3. Corollary: *Let* X,μ *be an* H*-space. Suppose* $\psi: X \to X_0 = K(QH^*(X,Z)/\text{torsion})$ *is such that* $QH^*(\psi,Z)/\text{torsion}$ *is an* *isomorphism. Then there exist a multiplication* m *for* X_0 *and a vector* $\bar\lambda = (\lambda_1, \lambda_2, \ldots)$ *of integers so that* $\Gamma(I_{\bar\lambda})\psi$ *is an* H-*map*.

Proof: Apply 4.4.2 for $\psi \times \psi\colon X \times X \to X_0 \times X_0$, $\psi\colon X \to X_0$ and $\mu\colon X \times X \to X$. Then there exists $\bar m\colon X_0 \times X_0 \to X_0$ and $\bar\lambda$ so that $\bar m[\Gamma(I_{\bar\lambda})\psi \times \Gamma(I_{\bar\lambda})\psi] \sim \Gamma(I_{\bar\lambda})\psi\mu$.

Consider the following:

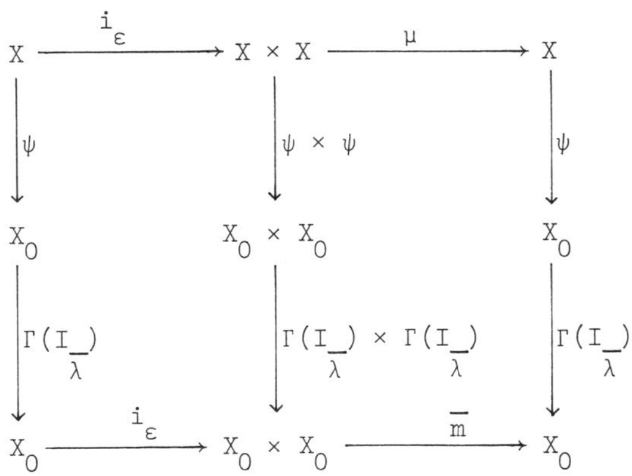

$\mu i_\varepsilon = 1$ $\varepsilon = 1,2$, imply $QH^*(\bar m i_\varepsilon, Z)/\text{torsion} = 1$ and consequently $\pi_*(\bar m i_\varepsilon) = 1$, $\bar m i_\varepsilon$ is a homotopy equivalence with a homotopy inverse γ_ε. $\Gamma(I_{\bar\lambda})\psi = \Gamma(I_{\bar\lambda})\psi\mu i_\varepsilon = \bar m i_\varepsilon \Gamma(I_{\bar\lambda})\psi$ implies $\gamma_\varepsilon \Gamma(I_{\bar\lambda})\psi = \Gamma(I_{\bar\lambda})\psi$. $m = \bar m(\gamma_1 \times \gamma_2)$ is an H-structure for X_0 and $\Gamma(I_{\bar\lambda})\psi$ is a μ-m H-map.

4.5. Mod $\mathbb{P}_1$ H-spaces

4.5.1. Definition: *A* mod $\mathbb{P}_1$ H-*space* is a pair X,μ where $\mu\colon X \times X \to X$ satisfies, $x \to \mu(x,*)$ and $x \to \mu(*,x)$ are $\mathbb{P}_1$ equivalences. (i.e. if $i_\varepsilon\colon X \to X \times X$ are the injections, $\varepsilon = 1,2$ then $h_\varepsilon = \mu i_\varepsilon$ are $\mathbb{P}_1$ equivalences).

4.5.2. Proposition: *Let* X *satisfy* (f.c.π). X *is a* mod $\mathbb{P}_1$ *H-space if and only if* X *is* $\mathbb{P}_1$*-equivalent to an* H*-space.*

Proof: If either $X \sim_{\mathbb{P}_1} X'$ with X' - an H-space or if X is a mod $\mathbb{P}_1$ H-space $H^*(X,Q)$ is a Hopf algebra. Hence, X is an H_0 space and $\mathbb{P}_1$ universal.

If X is $\mathbb{P}_1$ equivalent to an H-space X' (and we may assume X' satisfies (f.c.π) as well). 4.1.4 (b) implies the existence of $\mathbb{P}_1$ equivalences $h: X \xrightarrow{\approx \mathbb{P}_1} X'$ and $g: X' \xrightarrow{\approx \mathbb{P}_1} X$. If μ' is an H-structure for X' $g\mu'(h \times h)$ is a mod $\mathbb{P}_1$ H-structure for X.

Now suppose that X is a mod $\mathbb{P}_1$ H-space with a mod $\mathbb{P}_1$ H-structure $\mu: X \times X \to X$. Put $h_\varepsilon = \mu i_\varepsilon$, $i_\varepsilon: X \to X \times X$ $\varepsilon = 1,2$ - the injections. h_ε are $\mathbb{P}_1$ equivalences. Let $\psi: X \to X_0 = K(\pi_*(X)/\text{torsion})$ be any 0-equivalence. Using the methods of 4.4.2 we shall show that there exist a vector $\overline{\lambda} = (\lambda_1, \lambda_2, \ldots)$ and an H-structure $m: X_0 \times X_0 \to X_0$ so that

$$\Gamma(I_{\overline{\lambda}})\psi\mu \sim m[(\Gamma(I_{\overline{\lambda}})\psi h_1) \times (\Gamma(I_{\overline{\lambda}})\psi h_2)].$$

Again, this is a geometric realization of an algebraic observation: Let $G_* = QH^*(X,Z)/\text{torsion}$, $\chi: \phi(G_*) \to H^*(X,Z)$ the morphisms induced by ψ. Put $\alpha: H^*(X,Z) \otimes H^*(X,Z) \to H^*(X \times X,Z)$. One is looking for a vector $\overline{\lambda}$ and a morphism m^* so that the following diagram will commute:

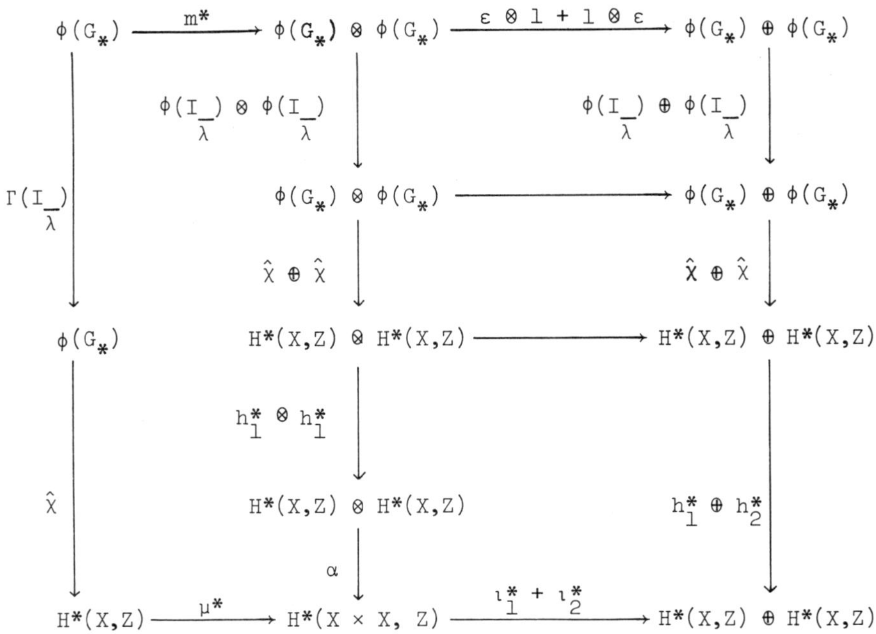

This can be seen by induction: Let $G_{\leq n} = \bigoplus_{m \leq n} G_m$. Suppose $\lambda_1,\ldots,\lambda_n$, m_n^* were found so that the diagram commutes with $G_{\leq n}$, $\bar\lambda_n = (\lambda_1,\ldots,\lambda_n)$, m_n^*, $\hat\chi_n\colon \phi(G_{\leq n}) \to H^*(X,Z)$ replacing G_*, $\bar\lambda$, m^* and $\hat\chi$ respectively. Now, if $\bar\mu^* = \mu^* - p_1^*h_1^* - p_2^*h_2^*$ then the image of $\bar\mu^*\chi_{n+1}\colon G_{n+1} \to H^{n+1}(X \times X, Z)$ is contained in $\ker i_1^* \oplus i_2^* = \operatorname{im}[H^*(X \wedge X, Z] \to \bar H^*(X \times X, Z)]$. As

$$\operatorname{im}[H^{n+1}(X \wedge X,Q) \to H^{n+1}(X \times X,Q)] \subset \{H^*(h_1,Q)(\hat\chi \otimes Q)(\phi(I_{\bar\lambda}) \otimes Q)[\phi(G_{\leq n}) \otimes Q]\} \otimes \{H^*(h_2,Q)(\hat\chi \otimes Q)(\phi(I_{\bar\lambda}) \otimes Q)[\phi(G_{\leq n}) \otimes Q]\}$$

there exists an integer λ_{n+1} so that

$$\operatorname{im} \lambda_{n+1}\bar\mu^*\hat\chi_{n+1} \subset \alpha\{h_1^*\hat\chi\phi(I_{\bar\lambda})[\phi(G_{\leq n})] \otimes h_2^*\chi\phi(I_{\bar\lambda})[\phi(G_{\leq n})]\}$$

and as G_{n+1} is free there exists $\bar m_{n+1}\colon G_{n+1} \to \phi(G_*) \otimes \phi(G_*)$ so that

$$\overline{\mu}^*\hat{x}_{n+1}\lambda_{n+1} \cdot 1 = \alpha(h_1^* \otimes h_2^*)(\hat{x} \otimes \hat{x})(\phi(I_{\overline{\lambda}_n}) \otimes \phi(I_{\overline{\lambda}_n}))\overline{m}_{n+1}.$$

Let $m^*_{n+1}\colon \phi(G_{\leq n+1}) \to \phi(G_{\leq n+1}) \otimes \phi(G_{\leq n+1})$ be induced by m^*_n and by

$\hat{m}_{n+1} = \eta \otimes 1 + 1 \otimes \eta + \overline{m}_{n+1}$ ($\eta\colon Z \to \phi(G_*)$ the unit).

Now that m and $\overline{\lambda}$ were defined one gets a geometric realization

$$\begin{array}{ccc}
X \times X & \xrightarrow{\ \mu\ } & X \\
\big\downarrow{\scriptstyle h_1 \times h_2} & & \\
X \times X & & \big\downarrow{\scriptstyle \psi} \\
\big\downarrow{\scriptstyle \psi \times \psi} & & \\
X_0 \times X_0 & & X_0 \\
\big\downarrow{\scriptstyle \Gamma(I_{\overline{\lambda}}) \times \Gamma(I_{\overline{\lambda}})} & & \big\downarrow{\scriptstyle \Gamma(I_{\overline{\lambda}})} \\
X_0 \times X_0 & \xrightarrow{\ m\ } & X_0
\end{array}$$

Put $\Gamma(I_{\overline{\lambda}})\psi = \psi_{\overline{\lambda}}$. As h_ε, $\varepsilon = 1,2$ are $\mathbb{P}_1$ equivalences

$X(\mathbb{P}_1, \psi_{\overline{\lambda}}) \sim X(\mathbb{P}_1, \psi_{\overline{\lambda}}h_\varepsilon)$ and by 4.3.1 one obtains a commutative diagram

$$\begin{array}{ccccc}
X & \xrightarrow{i_\varepsilon} & X \times X & \xrightarrow{\mu} & X \\
\downarrow{\scriptstyle h_\varepsilon} & & \downarrow{\scriptstyle h_1 \times h_2} & & \\
X & \xrightarrow{i_\varepsilon} & X \times X & & \downarrow{\scriptstyle \psi'_{\overline{\lambda}}} \\
\downarrow{\scriptstyle \psi'_{\overline{\lambda}}} & & \downarrow{\scriptstyle \psi'_{\overline{\lambda}} \times \psi'_{\overline{\lambda}}} & & \\
X(\mathbb{P}_1, \psi_{\overline{\lambda}}) & \xrightarrow{i_\varepsilon} & X(\mathbb{P}_1, \psi_{\overline{\lambda}}) \times X(\mathbb{P}_1, \psi_{\overline{\lambda}}) & \xrightarrow{\mu(\mathbb{P}_1)} & X(\mathbb{P}_1, \psi_{\overline{\lambda}}) \\
\downarrow{\scriptstyle \psi''_{\overline{\lambda}}} & & \downarrow{\scriptstyle \psi''_{\overline{\lambda}} \times \psi''_{\overline{\lambda}}} & & \downarrow{\scriptstyle \psi''_{\overline{\lambda}}} \\
X_0 & \xrightarrow{i_\varepsilon} & X_0 \times X_0 & \xrightarrow{m} & X_0
\end{array}$$

The uniqueness of 4.3.1 and the relations

$$\psi''_{\overline{\lambda}}\mu(\mathbb{P}_1)i_\varepsilon = mi_\varepsilon\psi''_{\overline{\lambda}} = \psi''_{\overline{\lambda}}$$

$$\mu(\mathbb{P}_1)i_\varepsilon\psi'_{\overline{\lambda}}h_\varepsilon = \psi'_{\overline{\lambda}}\mu i_\varepsilon = \psi'_{\overline{\lambda}}h_\varepsilon$$

imply $\mu(\mathbb{P}_1)i_\varepsilon = 1$, $\mu(\mathbb{P}_1)$ is an H-structure and $X \sim_{\mathbb{P}_1} X(\mathbb{P}_1, \psi_{\overline{\lambda}})$.

<u>4.5.2.1. Remark</u>: We make no minimality claims for the vector $\overline{\lambda}$ in the proof of 4.5.2 and one can apply the same method of construction if one is given a finite number of mod $\mathbb{P}_1$ H-structures μ_i for $X (i = 1, \dots k)$ and a fixed $\psi: X \xrightarrow{\approx 0} X_0$. If such a family is given one can choose the vector $\overline{\lambda}$ simultaneously for all i so that for every i there exists an H-structure m_i so that the following is commutative:

$$\begin{array}{ccc} X \times X & \xrightarrow{\ \mu_i\ } & X \\ \Big\downarrow h_1^{(i)} \times h_2^{(i)} & & \Big| \\ X \times X & & \Big| \\ \Big\downarrow \psi \times \psi & & \Big\downarrow \psi \\ X_0 \times X_0 & & X_0 \\ \Gamma(I_\lambda) \times \Gamma(I_\lambda) \Big\downarrow & & \Big\downarrow \Gamma(I_\lambda) \\ X_0 \times X_0 & \xrightarrow{\ m_i\ } & X_0 \end{array}$$

<u>4.5.3. Proposition</u>: *Let* X *satisfy* (f.c.). X *is an* H-*space if and only if* X *is a* mod p H-*space for every prime* p.

<u>4.5.3.1. Remark</u>: The early versions of this statement considered finite CW complexes. (See [Mislin]$_1$, Theorem 2.1). The proofs rely upon a preliminary announcement of Curjel (see [Curtis]$_1$, remark (e), p. 8-9) which was later retracted ([Curtis]$_1$, p. 1120). The corrected versions of the proof (e.g.: [Hilton, Mislin, Roitberg], Theorem III 1.8) had to assume some rational consistency condition of the various mod p H-structures of X. This rational restrictive assumption is removed in our proof.

As to Curjel's announcement that stated that every H-space with finite dimensional rational cohomology admits a multiplication inducing a co-associative (hence primitevly generated) rational cohomology Hopf algebra. As far as we know the validity of this statement for finite dimensional H-space is still an open question. However, if one considers H-spaces satisfying (f.c.π) and having finite dimensional rational cohomology, this statement is false. (See Stasheff's H-spaces problem 44, Neuchatel, Springer Lecture Notes 196, p. 134). This can be seen by observing the following example.

Let $Ht_{21}(SU) \times K(Z,23)$ be given a twisted multiplication μ_v, $v\colon Ht_{21}(SU) \wedge Ht_{21}(SU) \to K(Z,23)$

$$H^*(v,Z)\iota = u_5 \otimes u_5u_{13} - u_9 \otimes u_5u_9 + u_5u_{13} \otimes u_5 - u_5u_9 \otimes u_9$$

u_i a generator of $PH^i(Ht_{21}(SU), Z) \approx Z$. Let

$$x_{23},\ y_i \in H^*(Ht_{21}(SU) \times K(Z,23), Z_3)$$

be the mod 3 reduction of $1 \otimes \iota$ and of $u_i \otimes 1$ respectively. Then $z = P^1_3 x_{23} - y_5y_9y_{13}$ is primitive and if X is the fiber of $z \in [Ht_{21}(SU) \times K(Z,23),\ K(Z_3,27)]$, then X is an H-space, $\pi_n(X) = 0$ for $n > 26$ and $H^*(X,Q) = H^*(SU(12), Q)$.

The generator in $QH^{23}(X,Q)$ cannot be represented by a primitive regardless of the H-structure chosen for X. Indeed, if this generator is represented by a primitive, as $H^*(X,Z)$ is torsion free in $\dim < 23$, $\hat{x}_{23} \in H^{23}(X,Z_3)$ the image of x_{23} is primitive modulo a decomposable $\hat{d}$: $\hat{x}_{23} + \hat{d} \in PH^{23}(X,Z_3)$. This implies

$$P^1x_{23} + P^1\hat{d} = \hat{y}_5\hat{y}_9\hat{y}_{13} + P^1\hat{d} \in PH^{27}(X,23).$$

As there are no decomposable primitives of odd dimension in a Hopf algebra $\hat{y}_5\hat{y}_9\hat{y}_{13} + P^1\hat{d} = 0$. One can easily see that this relation must already hold in $H^*(Ht_{21}(SU), Z_3)$ (or even in $H^*(SU(11), Z_3)$). But

$w_5 \cdot w_9 \cdot w_{13} \notin \operatorname{im} P_3^1$, $0 \neq w_i \in PH^i(SU(11), Z_3)$.

As X satisfies (f.c.H) is an H-space if and only if $Ht_N(X)$ (N sufficiently large) is an H-space suffices to prove 4.5.3 for X satisfying (f.c.π). 4.5.3.2 carries the main technical burden of the proof of 4.5.3 and essentially shows that the rational consistency condition always holds for H-spaces satisfying (f.c.π).

<u>4.5.3.2. Proposition</u>: *Let* $X_1, \mu_1, X_2, \mu_2, X_0, m_0$ *be* H-*spaces*, X_i *satisfy* (f.c.π). *Suppose* X_0, m_0 *is* HA. *Let* $\psi_i\colon X_i \to X_0$ $i = 1,2$ *be* $\mathbb{P}$-$\mathbb{P}_i$ *equivalences*, $i = 1,2$, $\mathbb{P}_1 \cap \mathbb{P}_2 = \emptyset$. *Suppose* X_0 *admits* H-*structures* m_1, m_2 *so that* ψ_i *is a* $\mu_i - m_i$ H-*map*. *Then* X_i *and* X_0 *admit multiplications* $\tilde{\mu}_i$ *and* m_{12} *respectively so that* ψ_i *is a* $\tilde{\mu}_i - m_{12}$ H-*map*. *Moreover, for any finite subset* $\mathbb{P}_i'$ *of* $\mathbb{P}_i$ *and any* $n_0 > n$ $\tilde{\mu}_i$ *can be chosen so that:*

$$H^m(\mu_i, Z_p) = H^m(\tilde{\mu}_i, Z_p) \quad \textit{for} \quad p \in \mathbb{P}_i' \quad m < n_0$$

<u>Proof</u>: By 1.5.1 (a) the existence of m_i so that ψ_i is a $\mu_i - m_i$ H-map is equivalent to

$$HD(\psi_i, \mu_i, m_0) \in \operatorname{im}(\psi_i \wedge \psi_i)^*,$$

$$HD(\psi_i, \mu_i, m_0) = W_i(\psi_i \wedge \psi_i), \quad W_i\colon X_0 \wedge X_0 \to X_0$$

Suffices to find $\tilde{\mu}_i$ and $W\colon X_0 \wedge X_0 \to X_0$ so that for $i = 1,2$ $HD(\psi_i, \tilde{\mu}_i, m_0) = W(\psi_i \wedge \psi_i)$ for then $\tilde{\mu}_i$ and $m_{12} = W\Lambda + m_0$ will be the desired multiplications.

Let $\beta_n^i\colon X_i \to Ht_n(X_i)$ $i = 0,1,2$ be the Postnikov approximations of X_i and let $\gamma_{n+1}^i\colon X_i^{(n+1)} \to X_i$ be the fiber of $\beta_n^i(\gamma_{n+1}^i$ - the n-connective fibering).

Suppose inductively that multiplications $\tilde{\mu}_{i,n}: X_i \times X_i \to X_i$, $i = 1,2$, were constructed so that

$$HD(\psi_i, \tilde{\mu}_{i,n}, m_0) = W_{i,n}(\psi_i \wedge \psi_i) \quad \text{and}$$

$W_{1,n} - W_{2,n} = \gamma_n^0 u_n$, $u_n: X_0 \wedge X_0 \to X_0^{(n)}$. Now, $\psi_i^{(n)}: X_i^{(n)} \to X_0^{(n)}$ is a $\mathbb{P}$-$\mathbb{P}_i$ equivalence so there exists an integer λ_i prime to $\mathbb{P}$-$\mathbb{P}_i$ and a map $\Phi_i^{(n)}: X_0^{(n)} \to X_i^{(n)}$ so that $\psi_i^{(n)}\Phi_i^{(n)} = \lambda_i \cdot 1_{X_0^{(n)}}$. As $\mathbb{P}_1 \cap \mathbb{P}_2 = \emptyset$ λ_1 and λ_2 are relatively prime, and there exist integers a_1, a_2 (a_i prime to $\mathbb{P}$-$\mathbb{P}_i$) so that

$$-a_1\lambda_1 + a_2\lambda_2 = 1.$$

Let $\hat{v}_{i,n} = \hat{v}_i = \gamma_n^i \Phi_i^{(n)} a_i \cdot u_n$. Put $\tilde{\mu}_{i,n+1} = \hat{v}_i(\psi_i \wedge \psi_i)\Lambda + \tilde{\mu}_{i,n}$ (+ with respect to $\tilde{\mu}_{i,n}$). As in the proof of 1.5.1 (b) one has

$$\begin{aligned} HD(\psi_i, \tilde{\mu}_{i,n+1}, m_0)\Lambda = {} & HD(\psi_i, \tilde{\mu}_{i,n}, m_0)\Lambda[(\hat{v}_i\psi_i \wedge \psi_i\Lambda) \times \tilde{\mu}_{i,n}]\Delta_{X_i \times X_i} \\ & + \psi_i\hat{v}_i(\psi_i \wedge \psi_i)\Lambda + HD(\psi_i, \tilde{\mu}_{i,n}, m_0)\Lambda \end{aligned}$$

The first term can be rewritten as

$$\begin{aligned} & W_{i,n}(\psi_i \wedge \psi_i)[(\hat{v}_i(\psi_i \wedge \psi_i)\Lambda) \wedge \tilde{\mu}_{i,n}]\bar{\Delta}_{X_i \times X_i} = \\ & = W_{i,n}[(\psi_i\hat{v}_i(\psi_i \wedge \psi_i)\Lambda) \wedge \psi_i\tilde{\mu}_{i,n}]\bar{\Delta}_{X_i \times X_i} = \\ & = W_{i,n}\{[\psi_i\hat{v}_i\Lambda(\psi_i \times \psi_i)] \wedge [W_{i,n}(\psi_i \wedge \psi_i)\Lambda + m_0\psi_i \times \psi_i]\}\bar{\Delta}_{X_i \times X_i} = \\ & = W_{i,n}\{(\psi_i\hat{v}_i\Lambda) \wedge (W_{i,n}\Lambda + m_0)\}\bar{\Delta}_{X_i \times X_i}(\psi_i \times \psi_i) = \\ & = W_{i,n}\{\psi_i\hat{v}_i \wedge (W_{i,n}\Lambda + m_0)\}\alpha(\psi_i \wedge \psi_i)\Lambda \end{aligned}$$

where $\alpha: X_0 \wedge X_0 \to (X_0 \wedge X_0) \wedge (X_0 \times X_0)$ satisfies $\alpha\Lambda = (\Lambda \wedge 1)\bar{\Delta}_{X_0 \times X_0}$.

Put

$$W_{i,n+1} = W_{i,n}[\psi_i \hat{v}_i \wedge (W_{i,n}\Lambda + m_0)]\alpha + \psi_i \hat{v}_i + W_{i,n}$$

then

$$HD(\psi_i, \widetilde{\mu}_{i,n+1}, m_0) = W_{i,n+1}(\psi_i \wedge \psi_i).$$

Substitute $\hat{v}_i = \gamma_n^i \Phi_i^{(n)}(a_i \cdot u_n)$. The first term of $W_{i,n+1}$ decomposes as

$$\begin{array}{ccccc} X_0 \wedge X_0 & \xrightarrow{\alpha} & (X_0 \wedge X_0) \wedge (X_0 \times X_0) & \xrightarrow{a_i u_n \wedge (W_{i,n}\Lambda + m_0)} & X_0^{(n)} \wedge X_0 \\ & & & & \Big\downarrow {\scriptstyle W_{i,n}[\psi_i \gamma_n^i \Phi_i^{(n)} \wedge 1]} \\ & & & & X_0 \end{array}$$

As $X_0^{(n)} \wedge X_0$ is n-connected $[X_0^{(n)} \wedge X_0, Ht_n(X_0)] = 0$ and β_n^0 annihilates the first term:

$$\beta_n^0 W_{i,n+1} = \beta_n^0(\psi_i \hat{v}_i + W_{i,n}).$$

$$\psi_i \hat{v}_i = \psi_i \gamma_n^i \Phi_i^{(n)}(a_i u_n) = \gamma_n^0 \psi_i^{(n)} \Phi_i^{(n)}(a_i u_n) = \gamma_n^0(\lambda_i \cdot 1)(a_i \cdot u_n) =$$

$$= \gamma_n^0(\lambda_i a_i \cdot u_n) = (\lambda_i a_i) \cdot \gamma_n^0 u_n = \lambda_i a_i (W_{1,n} - W_{2,n})$$

$$\beta_n^0(W_{1,n+1} - W_{2,n+1}) = \beta_n^0[\lambda_1 a_1 (W_{1,n} - W_{2,n}) + W_{1,n} - W_{2,n}$$

$$- \lambda_2 a_2 (W_{1,n} - W_{2,n})] = (\lambda_1 a_1 + 1 - \lambda_2 a_2)\beta_n^0(W_{1,n} - W_{2,n}) = 0$$

It follows that $W_{1,n+1} - W_{2n+1} = \gamma_{n+1}^0 u_{n+1}$. As X_0 satisfies f.c.π for some N $X_0^{(N)} = *$, $u_N = 0$, $W_{1,N} = W_{2,N}$ and $HD(\psi_1, \widetilde{\mu}_{1,\mathbf{N}}, m_0) = HD(\psi_2, \widetilde{\mu}_{2,N}, m_0)$. Put $\widetilde{\mu}_i = \widetilde{\mu}_{i,N}$, $W = W_{i,N}$.

Now, λ_i can be replaced by $\lambda_i \cdot \hat{\lambda}_i$ where $\hat{\lambda}_i$ is any multiple of primes in $\mathbb{P}_i$, e.g. a high enough power of the product of primes in $\mathbb{P}_i' \subset \mathbb{P}_i$, $\mathbb{P}_i'$ finite. $\hat{v}_i = \hat{v}_{i,N}$ in the inductive step is then replaced by $\hat{\lambda}_i \cdot \hat{v}_i$ and if $\hat{\lambda}_i$ is properly chosen then in dim $< n_0$

$$H^*(\tilde{\mu}_{i,n+1}, Z_p) = H^*(\hat{\lambda}_i \hat{v}_i (\psi_i \wedge \psi_i)\Lambda + \tilde{\mu}_{i,n}, Z_p) = H^*(\mu_{i,n}, Z_p)$$

for $p \in \mathbb{P}_i'$ and $H^*(\tilde{\mu}_i, Z_p) = H^*(\mu_i, Z_p)$.

<u>4.5.3.3. Corollary</u>: *If* X *is both a* mod $\mathbb{P}_1$ *and a* mod $\mathbb{P}_2$ *H-space then* X *is a* mod $\mathbb{P}_1 \cup \mathbb{P}_2$ *H-space.*

<u>Proof</u>: Replace $\mathbb{P}_2$ by $\mathbb{P}_2 - \mathbb{P}_1$ if necessary so one may assume $\mathbb{P}_1 \cap \mathbb{P}_2 = \emptyset$. Let μ_i, $i = 1,2$ be mod $\mathbb{P}_i$ H-structures for X. Fix $\psi: X \xrightarrow{\approx} X_0 = K(QH^*(X,Z)/\text{torsion})$. By 4.5.2.1 there exist $\bar{\lambda} = (\lambda_1, \lambda_2, \ldots)$ and $m_i: X_0 \times X_0 \to X_0$ so that for $i = 1,2$ the following is commutative $(\psi_{\bar{\lambda}} = \Gamma(I_{\bar{\lambda}})\psi)$

$$\begin{array}{ccc}
X \times X & \xrightarrow{\mu_i} & X \\
\downarrow{\scriptstyle h_1^i \times h_2^i} & & \\
X \times X & & \downarrow{\scriptstyle \psi_{\bar{\lambda}}} \\
\downarrow{\scriptstyle \psi_{\bar{\lambda}} \times \psi_{\bar{\lambda}}} & & \\
X_0 \times X_0 & \xrightarrow{m_i} & X_0
\end{array}$$

By 4.5.2 $X(\mathbb{P}_i, \psi_{\bar{\lambda}})$ admits a multiplication $\mu(\mathbb{P}_i)$ so that $\psi''_{\bar{\lambda},\mathbb{P}_i}$: $X(\mathbb{P}_i, \psi_{\bar{\lambda}}) \to X_0$ are $\mu(\mathbb{P}_i)$, m_i H-maps. By 4.5.3.2 one obtains

$\tilde{\mu}_i$ and $m_{1,2}$ so that $\psi''_{\bar{\lambda},\mathbb{P}_i}$ are $\tilde{\mu}_i - m_{1,2}$ H-maps and consequently their pullback $X(\mathbb{P}_1 \cup \mathbb{P}_2, \psi_{\bar{\lambda}})$ is an H-space. $X \approx_{\mathbb{P}_1 \cup \mathbb{P}_2} X(\mathbb{P}_1 \cup \mathbb{P}_2, \psi_{\bar{\lambda}})$ is a mod $\mathbb{P}_1 \cup \mathbb{P}_2$ H-space.

Proof of 4.5.3: Suppose X is a mod p H-space for every p. Then X is an H_0 space. Let $\psi: X \xrightarrow{\approx 0} X_0 = K(QH^*(X,Z)/\text{torsion})$ then ψ is a $\mathbb{P}_1$ quivalence, $\mathbb{P}$-$\mathbb{P}_1$-finite. Hence X is a mod $\mathbb{P}_1$ H-space as well as mod $\mathbb{P}_i = \{p_i\}$ $i = 2,\ldots,k$ where $\{p_2,\ldots,p_k\} = \mathbb{P}\text{-}\mathbb{P}_1$. By a simple induction 4.5.3.3 implies that X is a mod $\mathbb{P}$ H-space - hence an H-space.

4.6. The genus of an H-space.

4.6.1. Definition: Let X be a universal space (i.e: $\mathbb{P}_1$ universal for every set $\mathbb{P}_1$). Suppose X satisfies (f.c.). *The genus of* X is the set

$$G(X) = \{[Y] \mid Y \sim_p X \text{ for every } p\}.$$

A property P of homotopy classes of spaces is said to be generic if $[X] \in P$ implies $G(X) \subset P$.

4.6.2. Example: By 4.5.3 being an H-space is generic (in the f.c. case).

Combining the results of [Zabrodsky]$_9$ and [Wilkerson] one gets a characterization of the genus of an H-space satisfying (f.c.):

4.6.3. Proposition: *Let* X *be an* H-*space satisfying* (f.c.π). *Then* $[Y] \in G(X)$ *if and only if there exists an* H-*space* X_0 *so that* $X \times X_0 \sim Y \times X_0$.

Let X be an H_0 space satisfying (f.c.). Fix an arbitrary 0-equivalence $\psi: X \to X_0 = K(QH^*(X,Z)/\text{torsion})$ and let t be an integer divisible by all torsion of $H^m(X,Z)$ $m \leq N(X)$ and by $\prod_i \exp \pi_i(\text{fiber } \psi)$. (For a finite abelian group A $\exp A$ is the smallest integer m satisfying $mA = 0$). Let ℓ be the number of integers n_i so that $QH^{n_i}(X,Z)/\text{torsion} \neq 0$. Finally, let

$$[X,X]_t = \{f: X \to X \mid f \text{ is a } p \text{ equivalence for every } p|t\}.$$

The main result of [Zabrodsky]$_9$ states:

<u>4.6.4. Proposition</u>: *Let* X *be an* H_0 *space satisfying* (f.c.). *Then* $G(X)$ *admits an abelian group structure with an exact sequence*

$$[X,X]_t \xrightarrow{\alpha} (Z_t^*/\pm 1)^\ell \xrightarrow{\xi} G(X) \to 0$$

where $\alpha(f) = (|\det QH^{n_1}(f,Z)/\text{torsion}|, \ldots, |\det QH^{n_\ell}(f,Z)/\text{torsion}|)$. (Note that although $[X,X]_t$ is not a group $\operatorname{im}\alpha$ is).

For the (f.c.π) case, ξ is given as follows:

<u>4.6.5.</u> Let $d_1,\ldots,d_\ell$ be integers, $(d_i,t) = 1$. Let

$$I_{d_1,\ldots,d_\ell}: QH^*(X,Z)/\text{torsion} \to QH^*(X,Z)/\text{torsion},$$

$I_{d_1,\ldots,d_\ell} \mid QH^{n_i}(X,Z)/\text{torsion} = d_i \cdot 1$. Then $\xi(d_1, d_2,\ldots,d_\ell)$ is the class of $\tilde{X}$ given by a pullback diagram

$$\begin{array}{ccc} \widetilde{X} & \longrightarrow & X \\ \downarrow & & \downarrow \psi \\ X_0 & \xrightarrow{\Gamma(I_{d_1,\ldots,d_\ell})} & X_0 \end{array}$$

4.6.6. If X satisfies (f.c.H) the same procedure can apply except that in this case t is chosen to be an integer divisible by the torsion primes of $H^*(X,Z)$ and by $\prod_{i \leq \dim X} \exp \pi_i(\text{fiber } \psi)$ and $\widetilde{X}$ is replaced by $H\ell_{\dim X}(\widetilde{X})$.

4.6.7. Example: *The genus of* $SU(n)$: Let X be an H_0 space satisfying (f.c.π). (It is obvious that if X satisfies (f.c.H) $G(X) \sim G(Ht_N(X))$ for N sufficiently large so our observations cover the (f.c.H) case as well).

One can choose an integer t that will satisfy the 4.6.4 hypothesis simultaneously for X and $Ht_n(X)$ for all n. As one has an obvious function $[X,X]_t \to [Ht_n(X), Ht_n(X)]_t$ and an epimorphism $(Z_t^*/\pm 1)^\ell \to (Z_t^*/\pm 1)^{\ell_n}$ (ℓ_n the number of integers $n_i \leq n$ with $QH^{n_i}(X,Z)/\text{torsion} \neq 0$) so that the 4.6.4 exact sequences can be compared

4.6.7.1

$$\begin{array}{ccccc} [X,X]_t & \xrightarrow{\alpha} & (Z_t^*/\pm 1)^\ell & \xrightarrow{\xi} & G(X) \to 0 \\ \downarrow{\scriptstyle Ht_n} & & \downarrow{\scriptstyle r} & & \downarrow{\scriptstyle h} \\ [Ht_n(X),Ht_n(X)]_t & \xrightarrow{\alpha_n} & (Z_t^*/\pm 1)^{\ell_n} & \xrightarrow{\xi_n} & G(Ht_n(X)) \to 0 \end{array}$$

Hence, $G(X) \to G(Ht_n(X))$ is an epimorphism. If we choose $X = SU_n = Ht_{2n}(SU)$ one **has** even more: $[SU_m, SU_m]_t \to [SU_k, SU_k]_t$ is an epimorphism. Indeed, suffices to show this for $k = m-1$: Let $f: SU_{m-1} \to SU_{m-1}$ be a $\mathbb{P}_t$ equivalence, t - the 4.6.4 integer for SU_n $n \geq m$, $\mathbb{P}_t = \{p \in P | \; p|t\}$. By a theorem of Bott ([Bott]$_2$)

$$\text{degree}[\pi_{2m-1}(SU) = Z \to PH_{2m-1}(SU) = Z] = (m-1)!$$

If $\psi: SU_n \to \prod_{m \leq n} K(Z, 2m-1)$ realizes $QH^*(SU_n, Z)/\text{torsion}$, $\pi_{2m}(\text{fiber } \psi) = Z_{m!}$, $m < n$, and $t = \prod_{m<n} m!$, $\mathbb{P}_t = \{p | \; p < n\}$. Furthermore, torsion $H^{2m}(SU_{m-1}, Z) = Z_{(m-1)!}$ and its generator is the k invariant of $SU_m \to SU_{m-1}$. f being a $\mathbb{P}_t$ equivalence yields a multiplication by λ of torsion $H^{2m}(SU_{m-1}, Z)$ $(\lambda, t) = 1$. f induces a diagram

$$\begin{array}{ccc}
SU_m & \xrightarrow{\hat{f}} & SU_m \\
\downarrow & & \downarrow \\
SU_{m-1} & \xrightarrow{f} & SU_{m-1} \\
\Big\downarrow k & & \Big\downarrow k \\
K(Z,2m) & \xrightarrow{\times \lambda} & k(Z,2m)
\end{array}$$

and $\hat{f} \in [SU_m, SU_m]_t$ with image $f \in [SU_{m-1}, SU_{m-1}]_t$.

Now 4.6.7.1 for $X = SU_m$, $n = 2m-3$ yields an epimorphism $h: G(SU_m) \to G(SU_{m-1})$ but as $Ht_{2m-3}: [SU_m, SU_m]_t \to [SU_{m-1}, SU_{m-1}]_t$ is onto $\ker h = \xi(\ker r)$, $\ker r = \{1, 1, \ldots, 1, d | \; d \in Z_t^*/\pm 1\}$. We shall show now that $(1, 1, \dots, d) \in \operatorname{im} \alpha$ if and only if $d \equiv \pm 1 \ ((m-1)!)$. Indeed, if $d \equiv \pm 1 \ ((m-1)!)$ one has

$$\begin{array}{ccccc}
SU(m-1) & \xrightarrow{1} & SU(m-1) & & \\
\downarrow & & \downarrow & & \\
SU(m) & \xrightarrow{f_d} & SU(m) & & \\
\downarrow w & & \downarrow w & & \\
S^{2m-1} & \xrightarrow{d} & S^{2m-1} & \xrightarrow{u} & BSU(m-1)
\end{array}$$

where u classifies the principal $SU(m-1)$ bundle

$$SU(m-1) \to SU(m) \to SU(m)/SU(m-1) = S^{2m-1}.$$

As order $u = (m-1)!$ $du = u$. Put $\tilde{f}_d = Ht_{2m-1}(f_d) = SU_m \to SU_m$ and $\alpha\tilde{f}_d = (1, \dots, 1, d)$. Conversely, suppose $f\colon SU_m \to SU_m$, $\alpha f = (1, 1, \dots, d)$. Put $f_0 = Ht_{2m-3}(f)\colon SU_{m-1} \to SU_{m-1}$. Then, $\alpha f_0 = 1, \dots, 1$ and f_0 is a homotopy equivalence and by [Zabrodsky]$_{10}$, Corollary 1.14: There exists $\hat{f}\colon SU_m \approx SU_m$ with $Ht_{2m-1}(\hat{f}) = f_0$. Hence, $\hat{f} - f = j\chi$, $j\colon K(Z, 2m-1) \to SU_m$ the fiber of $SU_m \to SU_{m-1}$, $\chi\colon SU_m \to K(Z, 2m-1)$. $\pi_k(j\chi) = 0$ for $k < 2m-1$, hence, by [Zabrodsky]$_{10}$ Corollary 1.8 degree $\pi_{2m-1}(j\chi)$ is divisible by

$$\text{degree}[\pi_{2m-1}(SU_m) \to PH_{2m-1}(SU_m)] = (m-1)!$$

As degree $\pi_{2m-1}(\hat{f}) = \pm 1$, degree $\pi_{2m-1}(f) = \pm 1 \pmod{(m-1)!}$. As d is the image of (degree $\pi_{2m-1}(f)$) in $Z_t^*/\pm 1$ $d \equiv 1 \bmod (m-1)!$.

It follows that $\ker[G(SU_m) \to G(SU_{m-1})] \approx Z^*_{(m-1)!}/\pm 1$ and one obtains a decomposition sequence

$$\begin{array}{ccccccc}
Z^*_{(n-1)!}/\pm 1 & & Z^*_{(n-2)!}/\pm 1 & & Z^*_2/\pm 1 & & \\
\downarrow & & \downarrow & & \downarrow & & \\
G(SU_n) & \longrightarrow & G(SU_{n-1}) & \to \dots \to & G(SU_3) & \to & G(SU_2) = 0
\end{array}$$

Consequently, $|G(SU_n)| = \prod_{m=2}^{n} \text{order } (Z^*_{(m-1)!}/\pm 1) = \prod_{m=2}^{n} [\varphi((m-1)!)/2]$

where φ is the Euler function.

$$|G(SU_k)| = 1 \quad k \leq 4, \qquad |G(SU_5)| = \varphi(24)/2 = 4$$

$$G(SU_6) = 4. \qquad \varphi(120)/2 = 4.16 = 64 \text{ etc.}$$

Finally, the epimorphism $G(SU(n)) \to G(SU_n)$ yields the estimate

$$|G(SU(n)| \geq \prod_{m \leq n} [\varphi(m-1)!/2].$$

We dare to conjecture that this inequality is indeed an equality.

4.7. Mixing homotopy types

4.7.1. Definition: Let X_i $i = 1,\dots,k$ be spaces with 0-equivalences $\psi_i: X_i \to X_0$ and let $\mathbb{P}_i$ $i = 1,\dots,k$ be a partition of the set of primes. *The mixing of* $X_i, \mathbb{P}_i, \psi_i$ is the space $\mathrm{Mix}(X_i, \mathbb{P}_i, \psi_i)$ obtained as the pullback of the maps $\psi_i'': X_i(\mathbb{P}_i, \psi_i) \to X_0$ $(X_i(\mathbb{P}_i, \psi_i)$ in the notations of 4.3).

4.7.2. Proposition: *Let* X_i *be universal spaces satisfying* (f.c.) *of the same type. Then*

4.7.2.1. $\mathrm{Mix}\ (X_i, \mathbb{P}_i, \psi_i) \sim_{\mathbb{P}_i} X_i$

4.7.2.2. *If for every* i X_i *is a* mod $\mathbb{P}_i$ H-*space then* $\mathrm{Mix}(X_i, \mathbb{P}_i, \psi_i)$ *is an* H-*space.*

4.7.2.3. *If there exists a space* X *so that for every* i $X_i \in G(X)$ *then* $\mathrm{Mix}\ (X_i, \mathbb{P}_i, \psi_i) \in G(X)$.

4.7.2.4. *If* $X_i = X$, $\psi_i = \psi$ *for all* i *then* $\mathrm{Mix}(X_i, \mathbb{P}_i, \psi_i) = X$.

Proof: 4.7.2.1 follows from the following observations:

If $f_i : X_i \xrightarrow{\sim\ \mathbb{P}-\mathbb{P}_i} X_0$ $i = 1,\ldots,k$, $\{\mathbb{P}_i\}$ disjoint sets of primes then the pullback of the f_i's is $\mathbb{P}_i$ equivalent to X_i. Indeed, by induction for $k = 2$

$$\begin{array}{ccc}
X_{12} & \xrightarrow[\mathbb{P}-\mathbb{P}_2]{g_2} & X_1 \\
{\scriptstyle \mathbb{P}-\mathbb{P}_1\sim}\Big\downarrow{\scriptstyle g_1} & & {\scriptstyle \mathbb{P}-\mathbb{P}_1\sim}\Big\downarrow{\scriptstyle f_1} \\
X_2 & \xrightarrow[\mathbb{P}-\mathbb{P}_2]{f_2} & X_0
\end{array}$$

$\mathbb{P}_1 \subset \mathbb{P} - \mathbb{P}_2$ implies $X_{12} \sim_{\mathbb{P}_1} X_2$, similarly $X_{12} \sim_{\mathbb{P}_2} X_1$, $f_2 g_1$ is a $\sim_{\mathbb{P}-\mathbb{P}_1 \cup \mathbb{P}_2}$. For $k > 2$, given $X_1,\ldots,X_k$, $f_1,\ldots,f_k$, $\mathbb{P}_1,\ldots,\mathbb{P}_k$ replace X_1,X_2, f_1,f_2, $\mathbb{P}_1,\mathbb{P}_2$ by X_{12} (the pullback of f_1 and f_2 as above), $f_{12} = f_2 g_1$, $\mathbb{P}_1 \cup \mathbb{P}_2$ and apply the induction for

$X_{12}, X_3,\dots,X_k, \mathbb{P}_1 \cup \mathbb{P}_2, \mathbb{P}_3,\dots,\mathbb{P}_k, f_{12}, f_3,\dots,f_k$: The pullback X of $f_{12}, f_3,\dots,f_k$ is the pullback of $f_1, f_2,\dots,f_k$ and $X \approx_{\mathbb{P}_i} X_i$, i 2, $X \approx_{\mathbb{P}_1 \cup \mathbb{P}_2} X_{12} \approx_{\mathbb{P}_i} X_i$ $i = 1,2$.

4.7.2.2 follows from 4.5.3 as $X \approx_{\mathbb{P}_i} X_i$ imply X a mod $\mathbb{P}_i$ H-space, hence mod p for all p. 4.7.2.3 is immediate from 4.7.2.1, 4.7.2.4 follows from 4.3.3.

The following relates the concepts of mixing and genus of an H_0 space:

<u>4.7.3. Proposition</u>: *Let* X *be an* H_0 *space satisfying* (f.c.). *Fix an arbitrary* 0-*equivalence* $\psi: X \to X_0 = K(\pi_*(X)/\text{torsion})$. *There exists an integer* λ_0 *so that if* $\psi_{\lambda_0} = (\lambda_0 1)\psi$ *is a* $\mathbb{P}_1$ *equivalence (in* $\dim \le N(X)$ *in the* (f.c.H) *case) then* $[Y] \in G(X)$ *if and only if there exists a* $\mathbb{P}_2 = \mathbb{P}-\mathbb{P}_1$ *equivalence* $\phi: X_0 \to X_0$ *of the form* $\phi = \Gamma(I_{\underline{\lambda}})$ *(in the notations of* 4.4*) so that*

$$Y \approx \text{Mix}(X, X; \mathbb{P}_1, \mathbb{P}_2; \phi\psi_{\lambda_0}, \psi_{\lambda_0})$$

<u>Proof</u>: Obviously $\text{Mix}(X, X; \mathbb{P}_1, \mathbb{P}_2; \phi\psi_{\lambda_0}, \psi_{\lambda_0}) \in G(X)$ for any 0-equivalences ψ, ϕ.

Conversely, suppose $[Y] \in G(X)$. As the (f.c.π) case is simpler we shall prove the (f.c.H) case.

As in 4.1.4 there exists an integer λ_0 so that $\lambda_0 1: X_0 \to X_0$ factors through $Ht_{N(X)}(S)$, where S is a product of odd dimensional spheres. $\psi_{\lambda_0} = \lambda_0\psi$ thus yields a commutative diagram

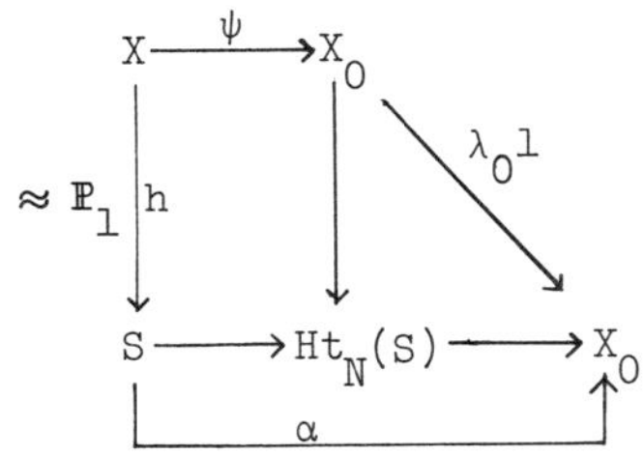

By 4.6.5 and 4.6.6 there exists $\phi = \Gamma(I_{\underline{\lambda}})$ so that Y is given as a homology approximation in a $\psi_{\lambda_0} - \phi$ pull back diagram. One can easily see that this is equivalent to the following:

4.7.3.0

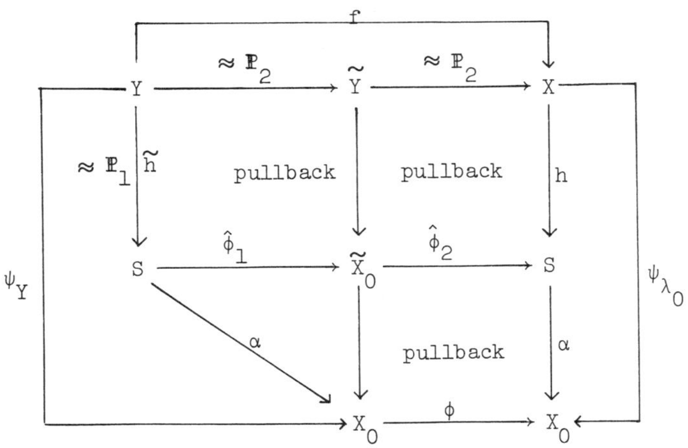

Consider now:

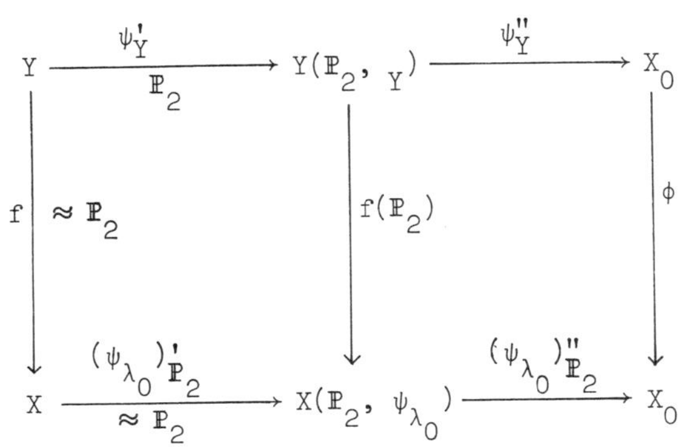

By the uniqueness of the 4.3.1 decomposition one obtains:

4.7.3.1

$$\begin{array}{ccccc}
Y & \xrightarrow{(\phi\psi_Y)'_{\mathbb{P}_2}} & Y(\mathbb{P}_2,\ \phi\psi_Y) & \xrightarrow{(\phi\psi_Y)''_{\mathbb{P}_2}} & X_0 \\
\| & & \downarrow \approx & & \| \\
Y & \xrightarrow{f(\mathbb{P}_2)\psi'_Y} & X(\mathbb{P}_2,\ \psi_{\lambda_0}) & \xrightarrow{(\psi_{\lambda_0})''_{\mathbb{P}_2}} & X_0
\end{array}$$

4.7.3.0 yields

$$\begin{array}{ccccccc}
Y & \underset{\approx\ \mathbb{P}_1}{\xrightarrow{\tilde{h}}} & \mathbf{S} & \xrightarrow{\alpha'} & S(\mathbb{P}_1,\alpha) & \underset{\approx\ \mathbb{P}_2}{\xrightarrow{\alpha''}} & X_0 \\
f \downarrow & & \downarrow \hat{\phi} = \hat{\phi}_2\hat{\phi}_1 & & \downarrow \hat{\phi}(\mathbb{P}_1) & & \downarrow \phi \\
X & \underset{\approx\ \mathbb{P}_1}{\xrightarrow{h}} & S & \xrightarrow{\alpha'} & S(\mathbb{P}_1,\alpha) & \underset{\approx\ \mathbb{P}_2}{\xrightarrow{\alpha''}} & X_0 \\
 & & & & & & \downarrow \phi \\
 & & & & & & X_0
\end{array}$$

Hence:

4.7.3.2

$$\begin{array}{ccccc}
X & \xrightarrow{(\phi\psi_{\lambda_0})'_{\mathbb{P}_1}} & X(\mathbb{P}_1,\phi\psi_{\lambda_0}) & \xrightarrow{(\phi\psi_{\lambda_0})''_{\mathbb{P}_1}} & X_0 \\
\| & & \downarrow \approx & & \| \\
X & \xrightarrow{\alpha' h} & S(\mathbb{P}_1,\ \alpha) & \xrightarrow{\phi\alpha''} & X_0
\end{array}$$

4.7.3.3

$$\begin{array}{ccccc}
Y & \xrightarrow{(\phi\psi_Y)_{\mathbb{P}_1}} & Y(\mathbb{P}_1,\ \phi\psi_Y) & \xrightarrow{(\phi\psi_Y)''_{\mathbb{P}_1}} & X_0 \\
\| & & \downarrow \approx & & \| \\
Y & \xrightarrow{\alpha'\tilde{h}} & S(\mathbb{P}_1,\alpha) & \xrightarrow{\phi\alpha''} & X_0
\end{array}$$

$Y = \mathrm{Mix}(Y,Y;\mathbb{P}_1,\mathbb{P}_2,\phi\psi_Y,\phi\psi_Y)$, hence, Y is the pull back of $(\phi\psi_Y)''_{\mathbb{P}_2}$ and

$(\Phi\psi_Y)''_{\mathbb{P}_1}$. By 4.7.3.1 one can replace $(\Phi\psi_Y)''_{\mathbb{P}_2}$ by $(\psi_{\lambda_0})''_{\mathbb{P}_2}$. By 4.7.3.3 $(\Phi\psi_Y)''_{\mathbb{P}_1}$ can be replaced by $\phi\alpha''$ which in turn can be replaced by $(\Phi\psi_{\lambda_0})''_{\mathbb{P}_1}$ (4.7.3.2). Hence, $Y = \mathrm{Mix}(X, X; \mathbb{P}_1, \mathbb{P}_2; \phi\psi_{\lambda_0}, \psi_{\lambda_0})$.

4.7.4. Corollary: *To be (of the homotopy type) of a finite CW loop space is generic, hence the mixing of finited CW loop spaces is a finite CW loop space.*

Proof: If $[Y] \in G(\Omega X)$ (ΩX satisfies (f.c.H)). By 4.7.3 $Y \approx \mathrm{Mix}(X,X; \Phi\psi, \psi; \mathbb{P}_1,\mathbb{P}_2)$. One can always choose $\psi\colon \Omega X \xrightarrow{\sim 0} \Omega X_0$ to be a loop map. $\phi = \Gamma(I_{\underline{\ }})$ and $\lambda_0 1$ are obviously loop maps and the procedure to construct $\mathrm{Mix}(\Omega X, \Omega X; \mathbb{P}_1, \mathbb{P}_2; \phi\psi_{\lambda_0}, \psi_{\lambda_0})$ is deloopable.

4.8. The non-classical H-spaces and other applications

Until 1968 the only known finite CW H-spaces were the classical ones: Those of the homotopy type of Lie groups, S^7 and their products.

In 1968 Hilton and Roitberg have discovered that E_{7W} - the S^3 bundle over S^7 classified by $7W \in \pi_6(S^3) \sim Z_{12}$ ($E_W = Sp(2)$) is an H-space and a non classical one. It was later shown by Stasheff ([Stasheff]$_3$) that E_{7W} is of the homotopy type of a finite dimensional loop space.

As $5 \equiv -7\ (12)$ $E_{5W} \sim E_{7W}$, $E_{5W} \sim_{\mathbb{P}_1} Sp(2)$ $\mathbb{P}_1 = \{p \mid p \neq 5\}$, $E_{7W} \sim_{\mathbb{P}_2} Sp(2)$, $\mathbb{P}_2 = \{p \mid p \neq 7\}$, hence, $E_{7W} \in G(Sp(2))$ and Stasheff's observation follows from 4.7.4. The latter is obviously a source for discovering an ∞ family of non classical finite dimensional loop spaces (e.g: By 4.6.7 $G(SU(n))$ in itself is a very rich source for such spaces). In fact till now we know of no finite dimensional loop space which is not in the genus of a Lie group. The same holds for mod 2 H-spaces where we know of no non-classical finite dimensional ones. The situation is different as far as general H-spaces are concerned. Adams (see [Adams]$_2$) was the first to observe that S^{2n+1} is a mod odd H-space. (Actually, from the point of view of 4.5.2 this follows from the much earlier observation that $2[\iota,\iota] = 0$, $\iota \in \pi_{2n+1}(S^{2n+1})$ the generator, and there exists a map $S^{2n+1} \times S^{2n+1} \to S^{2n+1}$ of type 2,1 which is a mod odd H-structure). This in turn yields a huge family of non classical H-spaces:

4.8.1. Let $\mathbb{P}_1$ be a set of odd primes, $X_{\mathbb{P}_1}$ a mod $\mathbb{P}_1$ H-space of type $n_1, n_2, \ldots, n_r$, $n_1 \leq n_2 \leq \ldots \leq n_r$ n_j-odd, i.e: $H^*(X_{\mathbb{P}_1}, Q) = \Lambda(x_{n_1}, \ldots, x_{n_r})$ $x_{n_i} \in H^{n_i}(X_{\mathbb{P}_1}, Q)$. There always exists a finite product of odd spheres $\tilde{S} = \prod_i S^{2m_i+1}$ so that $H^*(X_{\mathbb{P}_1} \times \tilde{S}, Q) \sim H^*(G,Q)$, G a Lie group. Let $\psi_1 \colon X_{\mathbb{P}_1} \times \tilde{S} \xrightarrow{\sim 0} X_0 = K(\pi_*(G)/\text{torsion})$ $\psi_2 \colon G \xrightarrow{\sim 0} X_0$. Then $\mathrm{Mix}(X_{\mathbb{P}_1} \times \tilde{S}, G; \mathbb{P}_1, \mathbb{P}-\mathbb{P}_1; \psi_1, \psi_2)$ is a non classical H-space.

4.8.1 illustrates how the construction of mod p H-spaces leads to the construction of non classical H-spaces. We shall present here a method to construct mod p H-spaces which is a general form of a construction described in [Stasheff]$_6$.

In the following proposition by a finite stable complex we mean a spectrum $\hat{K} = \{\Sigma^n K\}$ where K is a finite complex. If K is s-1 connected we write $H_i(\hat{K}) = H_{i+s}(K)$. $\dim \hat{K} = \dim K - s$.

<u>4.8.2. Proposition</u>: *Let* $\hat{K}$ *be a finite stable complex,* $H_{odd}(\hat{K},Z) = 0$ *Put* $D = \sum_m m\beta_m$, $r = \Sigma\beta_m$ *where* β_m *are the Betti numbers of* $\hat{K}$. *If* $1 < \ell$ *is an odd integer satisfying*

$$2(D + r\cdot\ell) < p\ell + p-2$$

then there exists an $\ell-1$ *connected finite* mod p *H-space* X *with a map* $\alpha: SX \to \hat{K}$ *of degree zero of the stable complex* $SX = \{\Sigma^n X\}$ *so that the composition*

$$H^*(\hat{K},Z_p) \xrightarrow{H^*(\alpha,Z_p)} H^*(SX,Z_p) = H^{*+\ell}(X,Z_p) \to QH^{*+\ell}(X,Z_p)$$

is an isomorphism.

<u>Proof</u>: Represent $\hat{K}$ by a $2N-1$ connected $< 4N$ dimensional complex K, N-large. The idea is to show that $\hat{X} = \Omega^{\bar{n}}\Sigma K$ for suitably chosen $\bar{n}$ will have the property $H\ell_k(\hat{X}) \sim_p H\ell_{2k}(\hat{X})$ for some k. The map $H\ell_k(\hat{X}) \times H\ell_k(\hat{X}) \to H\ell_{2k}(\hat{X})$ induced by Add: $\hat{X} \times \hat{X} \longrightarrow \hat{X}$ can be mod p deformed to a mod p H-structure $H\ell_k(\hat{X}) \times H\ell_k(\hat{X}) \to H\ell_k(\hat{X})$, $H\ell_k(\hat{X}) = X$.

The latter stage is quite simple, as $H\ell_m(\Omega Y)$ is p-universal for any space Y, given $h: H\ell_{2k}(\hat{X}) \xrightarrow{\approx p} H\ell_k(\hat{X})$ then in

$$\begin{array}{ccccc}
H\ell_k(\hat{X}) \times H\ell_k(\hat{X}) & \xrightarrow{\hat{m}} & H\ell_{2k}(\hat{X}) & \xrightarrow{h} & H\ _k(\hat{X}) \\
\downarrow & & \downarrow & & \\
\hat{X} \times \hat{X} & \xrightarrow{\text{Add}} & \hat{X} & &
\end{array}$$

$h\hat{m}$ is a mod p H-structure for $X = H\ell_k(X)$. The observation $H\ell_k(\hat{X}) \approx_p H\ell_{2k}(\hat{X})$ (depending on the choices of $\bar{n}$ and k) is a consequence of the algebraic observation $H_i(\hat{X},Z_p) = 0$ for $k < i \leq 2k$. To see this one has to analyse $H^*(\Omega^n\Sigma K, Z_p)$

Let Y be an H-space with $H^s(Y,Z_p) = 0$ for $s < 2m-1$. By [Browder]$_4$, theorem 5.14

(a) $\sigma^*: QH^s(Y,Z_p) \xrightarrow{\approx} PH^{s-1}(\Omega Y,Z_p)$ for $s < 2(m-1)p+1$.

A simple Hopf algebra argument yields

(b) $PH^{s-1}(\Omega Y,Z_p) \xrightarrow{\approx} QH^{s-1}(\Omega Y,Z_p)$ for $s < 2(m-1)p+1$.

Again by [Browder]$_6$, Theorem 5.14

(c) $QH^{s-1}(\Omega Y,Z_p) \xrightarrow{\approx} PH^{s-2}(\Omega^2 Y, Z_p)$ for $s < 2(m-1)p$

Finally

(d) $PH^{s-2}(\Omega^2 Y, Z_p) \xrightarrow{\approx} QH^{s-2}(\Omega^2 Y, Z_p)$ for $s < 2(m-1)p+2$

Putting (a) - (d) together one has

4.8.2.1 $\quad QH^s(Y,Z_p) \xrightarrow{\approx} QH^{s-2}(\Omega^2 Y, Z_p)$ for $s < 2(m-1)p$

$(H^{s-2}(\Omega^2 Y, Z_p) = 0$ for $s < 2m-1)$.

Consider $Y = \Omega\Sigma K$. Then

$$QH^s(Y,Z_p) \approx H^s(K,Z_p) \quad \text{for} \quad s < 4N,$$

$$H^s(Y,Z_p) = 0 \quad \text{for} \quad s < 2N,$$

$$H^{2s+1}(Y,Z_p) = 0.$$

As in (c) and (d) $QH^s(Y,Z_p) \xrightarrow{\approx} QH^{s-1}(\Omega Y, Z_p)$ for $s < 2Np$

Repeating 4.8.2.1 n times for ΩY one obtains

$H^s(K,Z_p) \approx QH^{s-1}(\Omega Y, Z_p) \approx QH^{s-2n-1}(\Omega^{2n+1} Y, Z_p)$ if $s - 2n-1 < 2(N-n)p-2$

and $s < 4N$ (or equivalently $s - 2n-1 < 4N - 2n-1$). Substitute $2s$ for for $s-2n$ one obtains

$$QH^{2s-1}(\Omega^{2n+1}Y, Z_p) = H^{2s+2n}(K,Z_p) = H^{2s+2n-2N}(\hat{K},Z_p)$$

for $2s-1 < 2(N-n)p-2$ and $2s-1 < 4N - 2n-1$. As $QH^{2s+1}(\hat{K},Z_p) = 0$ one obtains

$$QH^{2s}(\Omega^{2n+1}Y, Z_p) = 0 \quad \text{if} \quad s < (N-n)p-1 \quad \text{and} \quad s < 2N-n$$

Choose N and n so that:

$$2N+\ell > (\ell+1)p-2, \qquad \ell = 2(N-n) - 1$$

then $4N - 2n-1 = 2N + \ell > (\ell+1)p-2 = 2(N-n)p-2$, hence,

$$QH^{2s-1}(\Omega^{2n+1}Y, Z_p) = H^{2s+2n-2N}(\hat{K},Z_p) = H^{2s-\ell-1}(\hat{K},Z_p)$$

for $2s-1 < (\ell+1)p-2$. $QH^{2s}(\Omega^{2n+1}Y, Z_p) = 0$ for $2s < (\ell+1)p-2$.

Let A be the exterior algebra generated by $QH^{2s-1}(\Omega^{2n+1}Y, Z_p)$ $2s-1 < (\ell+1)p-2$. Then

$$k = \dim A = \sum_{2s-1<(\ell+1)p-2} (2s-1)\operatorname{rank} QH^{2s-1}(\Omega^{2n+1}Y, Z_p) =$$

$$= \Sigma(2s-1)\beta_{2s-\ell-1} = \Sigma(2s-\ell-1)\beta_{2s-\ell-1} + \ell\Sigma\beta_{2s-\ell-1} \leq D + \ell r$$

By the hypothesis $2k = 2 \dim A < p \cdot(\ell+1) - 2$ and consequently $A \sim H^*(\Omega^{2n+1}Y, Z_p)$ through dim $p(\ell+1) - 2$ and $X = H\ell_k(\Omega^{2n+1}Y) \sim_p H\ell_{2k}(\Omega^{2n+1}Y)$. One can easily argue that $D + \ell r < \frac{1}{2}(p(\ell+1) - 2)$ implies that $\dim \hat{K} < (\ell+1)p-2-\ell$ and hence

$$\alpha: \Sigma^{2n+2}X \to \Sigma K \quad \text{yields} \quad H^{*-\ell}(\hat{K},Z_p) \xrightarrow{\approx} QH^*(X,Z_p).$$

<u>4.8.3. Realizing Hopf algebras</u>: *Let* G *be a finite dimensional* H-*space. Suppose* $H^*(G,Z)$ *is odd torsion free. Given a finite set of odd primes* $\mathbb{P}_1$ *and Hopf algebra structures* ψ_p *for* $A_p = H^*(G,Z_p)$ $p \in \mathbb{P}_1$, *there exists an* H-*space* $\hat{G}$, *so that* $H^*(G,Z) = H^*(\hat{G},Z)$ *and for every* $p \in \mathbb{P}_1$ $H^*(\hat{G},Z_p) = A_p$ *as Hopf algebras.*

<u>Proof</u>: G is 0-equivalent to a product of odd spheres $\tilde{S} = \prod_{i=1}^{k} S^{2m_i+1}$. Let $\hat{S}_i$ be an H-space, mod $\mathbb{P}_1$ equivalent to S^{2m_i+1}. One can easily give $\hat{S} = \prod_{i=1}^{k} \hat{S}_i$ a multiplication $\hat{\mu}$ so that $H^*(\hat{S},Z_p) = A_p$ as a Hopf algebra, $p \in \mathbb{P}_1$. (This is being done by realizing the reduced coproduct $x_{2m_i+1} \to \psi_p\, x_{2m+1} - x_{2m+1} \otimes 1 - 1 \otimes x_{2m+1}$ via a map $(\prod_{j<i} \hat{S}_j) \wedge (\prod_{j<i} \hat{S}_j) \to \hat{S}_i$). One has 0-equivalences $\psi_1: \hat{S} \to K(\pi_*(G)/\text{torsion}) = G_0$, $\psi_2: G \to G_0$. By 4.5.3.2 one can find a multiplication $\hat{\mu}_1$ for $\hat{S}$ so that $H^*(\hat{\mu}_1,Z_p) = H^*(\hat{\mu},Z_p)$ $p \in \mathbb{P}_1$ and multiplications μ (for G) m (for G_0) so that ψ_1 is a $\hat{\mu}_1 - m$ H-map and ψ_2 a μ-m H-map.

$\mathrm{Mix}(\hat{S}, G; \mathbb{P}_1, \mathbb{P}-\mathbb{P}_1; \psi_1, \psi_2)$ is the desired H-space.

Chapter V

Non stable BP resolutions

Introduction

The Adams-Novikov spectral sequence proved to be effective in some computations in the stable homotopy groups of spheres (see [Zahler]). In this chapter we attempt to advocate for the use of the BP-spectrum as studied by Wilson (e.g. [Wilson] and subsequent papers) for some non stable problems.

The connection to H-spaces lies with the facts that our model spaces are H-spaces and some H-spaces invariants are occasionally used and that some of our applications deal with H-spaces.

We give here a very loose account of the prospective theory and our applications are just a scratch on the surface of possibilities and we dare say that already here its efficiency compare to the classical obstruction theory can be observed. (It does not imply however, that some ugly arithmetic computations can be always avoided).

Throughout this chapter we use many of Wilson's results without proof.

As we are to stay within the category of CW-complexes of finite type we adopt the Brown-Peterson original construction of the spectrum BP for our nonstable model. (See [Brown Peterson]). We use freely some of the results obtained there.

5.1. Killing homology p-torsion

5.1.1. Proposition: *Let* X *be a space. For a given prime* p *there exists a space* $F(X) = F_{(p)}(X)$ *and a map* $h: F(X) \to X$ *so that:*

(a) $H_*(F(X), Z)$ *is* p-*torsion free.*

(b) $\pi_*(\text{fiber } h)$ *is torsion free.*

(c) There exists a commutative diagram

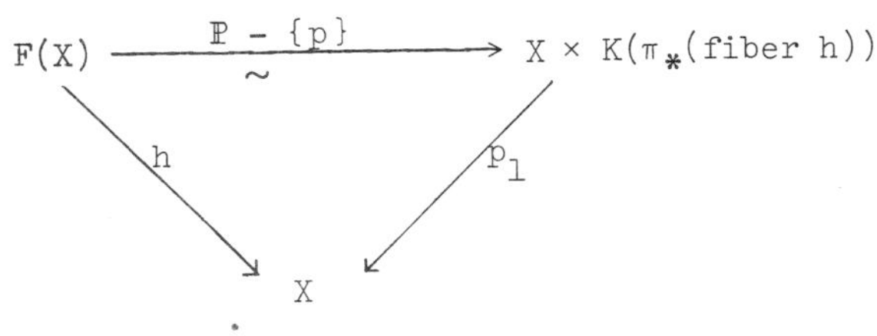

(d) $h: F(X) \to X$ *does not dominate* mod p *any non trivial space over* X. *i.e. If in the following commutative diagram*

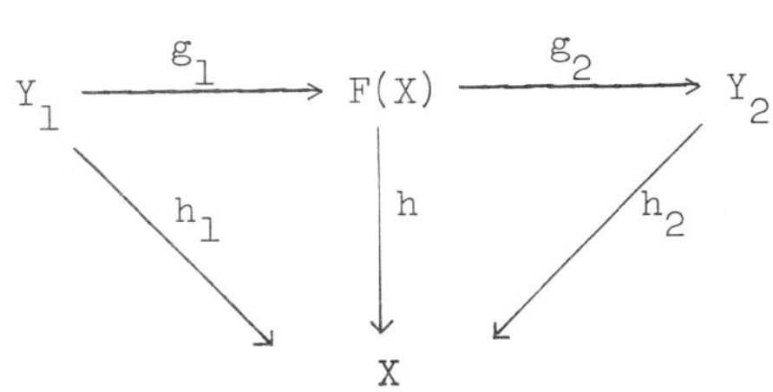

$g_2 g_1$ *is a* mod p *equivalence, then so are* g_i.

Proof: $F(X)$ is constructed inductively as follows: Suppose $h_n: F_n(X) \to X$ is constructed so that $\pi_m(\text{fiber } h_n)$ is 0 for $m \geq n$ and torsion free for all m, $H_m(F_n(X), Z)$ is p-torsion free for $m \leq n$ and one has a commutative diagram:

$$\begin{array}{ccc} F_n(X) & \xrightarrow{\approx\ \mathbb{P}\text{-}p} & X \times K(\pi_*(\text{fiber } h_n)) \\ & {}_{h_n}\searrow \quad \swarrow{}_{\mathbb{P}_1} & \\ & X & \end{array}$$

Let $\tilde{k}_n : F_n(X) \to K(H_{n+1}(F_n(X), Z), n+1)$ yield an isomorphism on homology in dim n+1. Put $\tilde{A}_{n+1} = H_{n+1}(F_n(X), Z)/ \neq p$ torsion, $k_n : F_n(X) \to K(\tilde{A}_{n+1}, n+1)$ induced by $\tilde{k}_n$ and the quotient. Let A_{n+1} be a free abelian group, $\alpha_{n+1}: A_{n+1} \to \tilde{A}_{n+1}$ a surjection and $\alpha_{n+1} \otimes Z_p$ an isomorphism. Let $F_{n+1}(X)$ complete the following into a pullback diagram:

$$\begin{CD} F_{n+1}(X) @>{u_{n+1}}>> K(A_{n+1}, n+1) \\ @V{h_{n+1,n}}VV @VV{K(\alpha_{n+1}, n+1)}V \\ F_n(X) @>{k_n}>> K(\tilde{A}_{n+1}, n+1) \end{CD}$$

Using the Serre exact sequence one can see (mod $\neq$ p torsion) that $H_{n+1}(\text{fiber } k_n, Z) = 0$ and $H_{n+1}(u_{n+1}, Z)$ is an isomorphism. As $H_m(h_{n+1,n}, Z)$ is an isomorphism for $m \leq n$ $H_m(F_{n+1}(X), Z)$ is p-torsion free for $m \leq n+1$.

Put $h_{n+1} = h_n h_{n+1,n}$ and one has a fibration

$$\text{Fiber } h_{n+1,n} \to \text{Fiber } h_{n+1} \to \text{Fiber } h_n .$$

Fiber $h_{n+1,n}$ = Fiber $K(\alpha_{n+1}, n+1) = K(\ker \alpha_{n+1}, n+1)$. $\ker \alpha_{n+1} \subset A_{n+1}$ is free, hence, $\pi_m(\text{fiber } h_{n+1})$ is free: $\pi_m(\text{fiber } h_{n+1}) = \pi_m(\text{fiber } h_n)$ $m \leq n$, $\pi_{n+1}(\text{fiber } h_{n+1}) = \pi_{n+1}(\text{fiber } h_{n+1,n})$.

$F_{n+1}(X)$ can be constructed alternatively by composing

$$F_n(X) \xrightarrow[k_n]{} K(\widetilde{A}_{n+1}, n+1) \xrightarrow[\gamma_n]{} K({}^pH_{n+1}(F_n(X), Z), n+1)$$

(γ_n induced by any splitting $\widetilde{A}_{n+1} \to {}^pH_{n+1}$.) Let $\hat{\alpha}_{n+1}: \hat{A}_{n+1} \to {}^pH_{n+1}((F_n(X), Z), n+1)$ be any epimorphism of a free group $\hat{A}_{n+1}$ so that $\hat{\alpha}_{n+1} \otimes Z_p$ is an isomorphism. $F_{n+1}(X)$ is then the pullback

$$\begin{array}{ccc} F_{n+1}(X) & \xrightarrow{\hat{u}} & K(\hat{A}_{n+1}, n+1) \\ \downarrow{\scriptstyle h_{n+1,n}} & & \downarrow{\scriptstyle K(\hat{\alpha}_{n+1}, n+1)} \\ F_n(X) & \xrightarrow{\gamma_n k_n} & K({}^pH_{n+1}(F_n(X), Z), n+1) \end{array}$$

$F_{n+1}(X)$ is the fiber of

$$\gamma_n k_n - K(\hat{\alpha}_{n+1}, n+1): F_n(X) \times K(\hat{A}_{n+1}, n+1) \to K({}^pH_{n+1}(F_n(X),Z),n+1)$$

hence, $F_{n+1}(X) \approx_{\mathbb{P}-p} F_n(X) \times K(\hat{A}_{n+1}, n+1)$. Now $\hat{A}_{n+1} = \pi_{n+1}(\text{fiber } h_{n+1,n})$

$$F_{n+1}(X) \xrightarrow{\approx\ \mathbb{P}-p} F(X) \times K(A_{n+1},n+1) \xrightarrow{\approx\ \mathbb{P}-p} X \times K(\pi_*(\text{fiber } h_n) \times K(\hat{A}_{n+1},n+1) =$$

$= X \times K(\pi_*(\text{fiber } h_{n+1}))$ The following commutes:

$$\begin{array}{ccc} F_{n+1}(X) & \xrightarrow{\mathbb{P}\text{-}p} & X \times K(\pi_*(\text{fiber } h_{n+1})) \\ & {\scriptstyle h_{n+1}} \searrow \quad \swarrow {\scriptstyle p_1} & \\ & X & \end{array}$$

$F(X)$, $h = \varprojlim F_n(X)$, $\varprojlim h_n$ satisfy (a) - (c).

To see (d), suppose one is given

$$\begin{array}{ccccc} Y_1 & \xrightarrow{g_1} & F(x) & \xrightarrow{g_2} & Y_2 \\ & {\scriptstyle h_1} \searrow & \downarrow {\scriptstyle h} & \swarrow {\scriptstyle h_2} & \\ & & X & & \end{array}$$

$g_2 g_1$ a p-equivalence. Decompose h, h_i into a Postnikov decomposition:

$$\begin{array}{ccccc} Y_{1,n} & \xrightarrow{g_{1,n}} & F_n(X) & \xrightarrow{g_{2,n}} & Y_{2,n} \\ \downarrow {\scriptstyle h_{1,n,n-1}} & & \downarrow {\scriptstyle h_{n,n-1}} & & \downarrow {\scriptstyle h_{2,n,n-1}} \\ Y_{1,n-1} & \xrightarrow{g_{1,n-1}} & F_{n-1}(X) & \xrightarrow{g_2,n-1} & Y_{2,n-1} \end{array}$$

$H_n(h_{i,n,n-1}, Z)$ is surjective (fiber $h_{i,n,n-1}$ being $n-1$ connected) and if one supposes inductively that $g_{i,n-1}$ are p-equivalences

$H_n(g_{1,n-1})$: $H_n(Y_{1,n-1}, Z) \xrightarrow{\approx p} H_n(F_{n-1}(X), Z)$, $H_n(h_{n,n-1}, Z) \otimes Z_p$ an isomorphism, hence, $H_n(g_{1,n}, Z) \otimes Z_p$ is a surjection and $H_n(g_{1,n}, Z)$ is a p-surjection. Having a mod p left inverse $H_n(g_{2,n}, Z)$,

$H_n(g_{1,n}, Z)$ is a mod p isomorphism. As

$$0 \to H_n(\text{fiber } h_{n,n-1}, Z) \to H_n(F_n(X), Z) \to H_n(F_{n-1}(X), Z) \to 0$$

is mod p exact it follows that $H_n(\text{fiber } h_{1,n,n-1}, Z) \to H_n(\text{fiber } h_{n,n-1}, Z)$ is mod p surjective. Again, having a mod p left inverse

$$H_n(\text{fiber } h_{n,n-1}, Z) \to H_n(\text{fiber } h_{2,n,n-1}, Z)$$

$$H_n(\text{fiber } h_{1,n,n-1}, Z) \xrightarrow{\sim p} H_n(\text{fiber } h_{n,n-1}, Z).$$

As fiber $h_{1,n,n-1} = K(\pi_n(\text{fiber } h_1), n)$, fiber $h_{n,n-1} = K(\pi_n(\text{fiber } h), n)$ it follows that $g_{1,n}$ yields a mod p - equivalence fiber $h_{1,n,n-1} \to$ fiber $h_{n,n-1}$ and $g_{1,n}$, $g_{2,n}$ are p-equivalences.

<u>5.1.1.1. Remark</u>: As far as the mod p type of $h: F(X) \to X$ is concerned we can find models with either $\pi_n(\text{fiber } h)$ or $H_n(F(X), Z)$ torsion free but not both, otherwise the mod p minimality (property (d)) will be lost. The mod p invariant version of property (c) is the weaker

$$\begin{array}{ccc} F(X) & \xrightarrow{\approx 0} & X \times K(\ _*(\text{fiber } h)) \\ & {\scriptstyle h} \searrow \quad \swarrow {\scriptstyle p_1} & \\ & X & \end{array}$$

The construction of $F(X)$ is not unique. However, once a 5.1.1 construction $h: F(X) \to X$ is chosen one obtains invariants $\alpha_n: H_n(F_n(X), Z)/\neq p \text{ torsion} = A_n \to \tilde{A}_n$.

5.1.2. Proposition:

(a) *Let* $H^*(Y,Z)$ *be* p-*torsion free. Then any map* $f\colon Y \to X$ *lifts to a map* $\hat{f}\colon Y \to F(X)$. *Moreover, if for*
$\alpha_{n*}\colon H^n(Y, A_n) \longrightarrow H^n(Y, \tilde{A}_n)$ *(which is onto)*
one chooses a left inverse (as a function of sets)
$J_n\colon H^n(Y,\tilde{A}_n) \longrightarrow H^n(Y, A_n)$, $\alpha_{n*}J_n = 1$ *then* $\hat{f}$ *can be chosen in a unique way so that* $[u_n\hat{f}_n] = J_n[k_{n-1}\hat{f}_{n-1}]$
$(\hat{f}_m\colon Y \to F_m(X))$.

(b) *If* $g\colon X \xrightarrow{\sim p} \tilde{X}$ *then any map* $\hat{g}\colon F(X) \to F(\tilde{X})$ *(lifting* $gh\colon F(X) \to X \to \tilde{X}$) *is a* p *equivalence.*

(c) *If* X *is a* mod p H-*space then so is* F(X) *and* h *is a* mod p H-*map.*

Proof: (a) Suppose J is given. Suppose inductively f lifts to $F_n(X)$. One has the following commutative diagram

$$\begin{array}{ccc} Y & \xrightarrow{J_{n+1}[k_n\hat{f}_n]} & K(A_{n+1}, n+1) \\ \Big\downarrow \hat{f}_n & & \Big\downarrow K(\alpha_{n+1}, n+1) \\ F_n(X) & \xrightarrow{\quad k_n \quad} & K(\tilde{A}_{n+1}, n+1) \end{array}$$

which yields a map $\hat{f}_{n+1}\colon Y \to F_{n+1}(X)$, $(F_{n+1}(X)$ - the pullback of k_n and $K(\alpha_{n+1}, n+1))$:

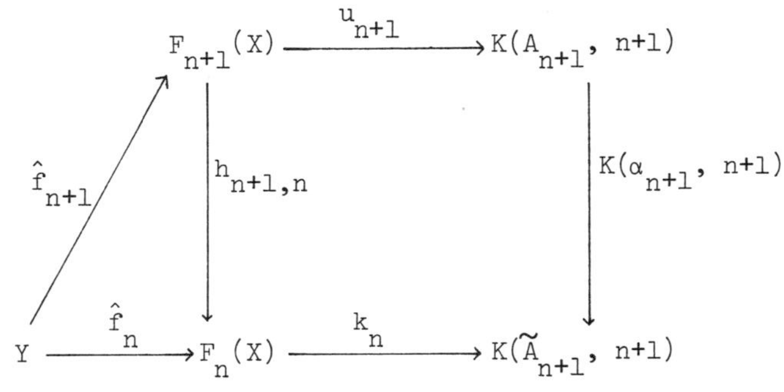

Now any lifting $\hat{f}'_{n+1}: Y \to F_{n+1}(X)$ of $\hat{f}_n$ induced by $J: [u_{n+1}\hat{f}'_{n+1}] = J_{n+1}[k_n\hat{f}_n]$ will satisfy $[u_{n+1}\hat{f}'_{n+1}] = J_{n+1}[k_n\hat{f}_n] = [u_{n+1}\hat{f}_{n+1}]$.

Hence, if $\hat{f}_{n+1} = (\hat{f}_n \times \gamma \times u_{n+1}\hat{f}_{n+1})\Delta_Y^2$, $\gamma: Y \to PK(\tilde{A}_{n+1}, n+1)$ then $\hat{f}'_{n+1} = (\hat{f}_n \times \lambda\cdot\gamma \times u_{n+1}\hat{f}_{n+1})\Delta_Y^2$ $\lambda: Y \to \Omega K(\tilde{A}_{n+1}, n+1)$, $\lambda\cdot\gamma$ - the pointwise multiplication of functions. As $H^{n+1}(Y,Z)$ is p-torsion free any map $\lambda: Y \to \Omega K(\tilde{A}_{n+1}, n+1)$ lifts to $\hat{\lambda}: Y \to \Omega K(A_{n+1}, n+1)$ and $\hat{\lambda}\cdot(u_{n+1}\hat{f}_{n+1})$ represents a self homotopy of $u_{n+1}\hat{f}_{n+1}$ which induces a homotopy corel $h_{n+1,n}$ of $\hat{f}_{n+1}$ and $\hat{f}'_{n+1}$.

(b) Suppose inductively $F_n(X) \xrightarrow[\sim p]{\hat{g}_n} F_n(X')$. Given the map of pullbacks:

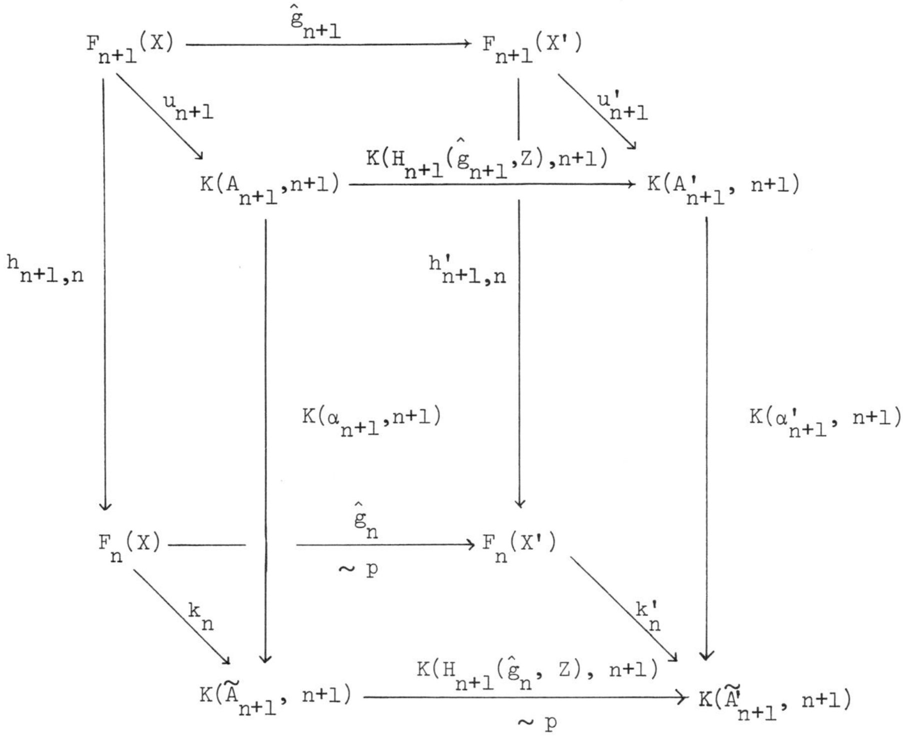

As $\alpha_{n+1} \otimes Z_p$ and $\alpha'_{n+1} \otimes Z_p$ are isomorphisms $H_{n+1}(\hat{g}_{n+1}, Z)$ is a p-isomorphism. Consequently $K(H_{n+1}(\hat{g}_{n+1}, Z), n+1)$ is a p equivalence and so is $\hat{g}_{n+1}$.

(c) Let $\mu: X \times X \to X$ be a mod p H-structure. By (a) one has

$\varepsilon = 1,2$

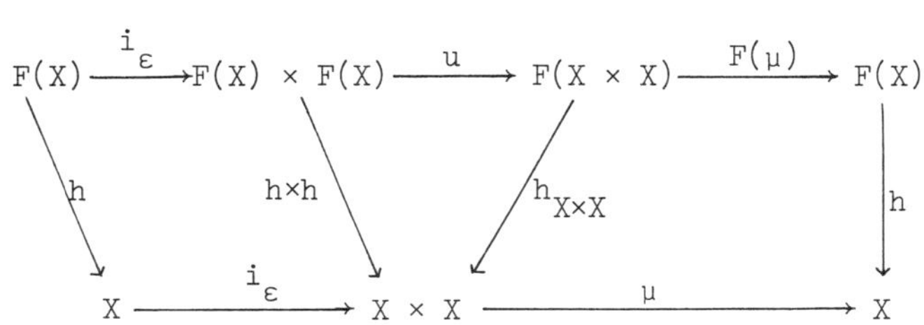

As μi_ε is a p-equivalence, hence, by (b) $F(\mu)ui_\varepsilon$ is a p-equivalence, $F(\mu)u$ is a mod p H-structure for $F(X)$ and h is a mod p H-map.

<u>5.1.3. Proposition</u>: $h: F(X) \to X$ *is characterized by properties* (a) - (d) *of* 5.1.1 *i.e.: If* $\hat{h}: Y \to X$ *satisfy* (a) - (d) *with* Y, $\hat{h}$ *replacing* $F(X)$, h *then there exists a* mod p *equivalence* $g: Y \xrightarrow{\sim p} F(X)$ *and* $hg \sim \hat{h}$.

<u>Proof</u>: 5.1.1 (a) and 5.1.2 (a) imply the existence of g, $hg \sim \hat{h}$. Now, 5.1.1 (b) and (c) for Y imply that the k invariants of $Y \to X$ are all integral of order power of p and the fact that $H_*(F(X), Z)$ is p-torsion free implies the existence of $v: F(X) \to Y$ $\hat{h}v \sim h$. Considering

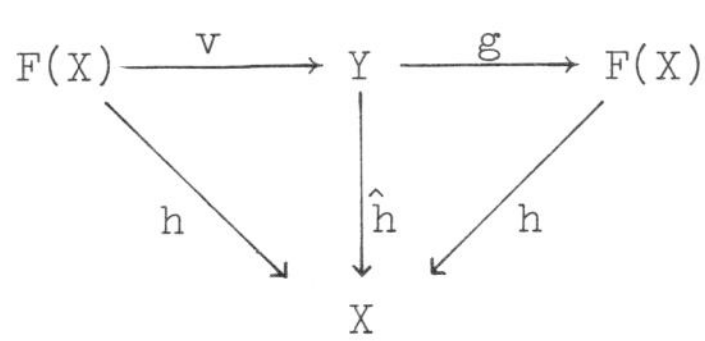

by 5.1.2 (b) gv is a p-equivalence and by property 5.1.1 (d) for Y both v and g are p-equivalences.

<u>5.2. Wilson's B(n,p)'s</u>

<u>5.2.1. Definition</u>: Let n be an integer and p - a prime. Wilson's $B(n,p)$ is the space $B(n,p) = F_{(p)}(K(Z,n)) = F_{(p)}(K(Z_p,n))$. ($B(n,p)$ corresponds to Y_n in [Wilson]) 5.1.1 and 5.1.3 yield the following mod p characterization of $B(n,p)$:

<u>5.2.2. Proposition</u>: *The* mod p *type of* B(n,p) *is characterized by:*

(a) B(n,p) *is* n-1 *connected* mod p, *i.e.:* $\pi_m(B(n,p)) \otimes Z_p = 0$ *for* $m < n$, $\pi_n(B(n,p)) \otimes Z_p \neq 0$

(b) $H^*(B(n,p), Z)$ *is* p-*torsion free.*

(c) $\pi_*(B(n,p))$ *is* p-*torsion free.*

(d) B(n,p) *is an* H_0 *space.*

(e) B(n,p) *does not dominate* mod p *any non trivial space, i.e.: If in*

$$Y_1 \xrightarrow{g_1} B(n,p) \xrightarrow{g_2} Y_2$$

$g_2 g_1$ *is a* mod p *equivalence and* $\widetilde{H}_*(Y_1, Z_p) \neq 0$ *then* g_i *are* mod p *equivalences.*

<u>Proof</u>: We first show that B(n,p) satisfies (a) - (e). By the definition of $F_{(p)}$ (5.1.1) (a) holds for B(n,p). By 5.1.1 (a), B(n,p) satisfies 5.2.2 (b), by 5.1.1 (b) and (c) B(n,p) satisfies 5.2.2 (c) and (d). By 5.1.2 (c) B(n,p) is a mod p H-space so if

$Y_1 \xrightarrow{g_1} B(n,p) \xrightarrow{g_2} Y_2$ $g_2 g_1$ - mod p equivalence $B(n,p) \approx_p Y_1 \times \hat{B}$.

As $\pi_n(B(n,p))/(\neq p)\text{torsion} = Z$ either $\pi_n(Y_1)/(\neq p)\text{torsion} = Z$ or $\pi_n(\hat{B})/(\neq p)\text{torsion} = Z$. With no loss of generality assume the latter. Then there exists

$$\begin{array}{ccccc} \hat{B} & \xrightarrow{i_2} & B(n,p) \approx_p Y_1 \times \hat{B} & \xrightarrow{p_2} & \hat{B} \\ & \searrow & \downarrow & \swarrow & \\ & & K(Z,n) & & \end{array}$$

By 5.1.1 (d) i_2, p_2 are p equivalences and $\tilde{H}_*(Y_1, Z_p) = 0$. The alternative assumption $\pi_n(Y_1)/(\neq p)\text{torsion} = Z$ will imply $Y_1 \approx_p B(n,p)$.

Conversely, given Y satisfying (a) - (e). Then $\pi_n(Y)$ is p-torsion free and let $\hat{h}: Y \to K(\pi_n(Y)/(\neq p)\text{torsion}, n)$ a mod p isomorphism of π_n. Let $Z \xrightarrow{\alpha_1} \pi_n(Y)/{\neq}p \text{ torsion} \xrightarrow{\alpha_2} Z$ $\alpha_2 \cdot \alpha_1 = 1$. One obtains

$$\begin{array}{ccccc}
B(n,p) & \xrightarrow{u} & Y & \xrightarrow{v} & B(n,p) \\
\downarrow{\scriptstyle h} & & \downarrow{\scriptstyle \hat{h}} & & \downarrow{\scriptstyle h} \\
K(Z,n) & \xrightarrow{(\alpha_1)_*} & K(\pi_n(Y)/(\neq p)\text{torsion}, n) & \xrightarrow{(\alpha_2)_*} & K(Z,n)
\end{array}$$

Indeed $(\alpha_2)_*\hat{h}$ lifts by 5.1.2 (a) and 5.2.2 (b). By 5.2.2 (c) and (d) the k invariants of $\hat{h}$ are integral of finite order and we may assume they are of order p^r, as $H_*(B(n,p), Z)$ is p-torsion free $(\alpha_1)_*h$ lifts. Now $(\alpha_2)_*(\alpha_1)_*$ is a homotopy equivalence so by 5.1.2 (b) vu is a mod p equivalence and by property (e) for Y u is a mod p equivalence (and obviously so are $(\alpha_i)_*$).

<u>5.2.3. Proposition</u>: *Let* X *be an* H_0 *space with* $H_*(X,Z)$ *and* $\pi_*(X)$ p-*torsion free. Then* $X \approx_p \prod_i B(n_i, p)$.

<u>Proof</u>: First note that being an H_0 space and having p-torsion free homotopy groups is equivalent (up to mod p equivalence) to the fact that the k invariants of X are integral of order power of p.

We first show that X is a mod p H-space. Suppose inductively that the following commutes:

$$\begin{array}{ccccc}
X \vee X & \xrightarrow{F} & X & \xrightarrow{h_n} & Ht_n X \\
\downarrow{\scriptstyle i} & & & \nearrow & \downarrow{\scriptstyle h_{n,n-1}} \\
X \times X & & \xrightarrow{m_{n-1}} & & Ht_{n-1} X \\
\downarrow{\scriptstyle \Lambda} & & & & \downarrow{\scriptstyle k_n} \\
X \wedge X & & \xrightarrow{\hat{m}_{n-1}} & & K(\pi_n X,\ n+1)
\end{array}$$

The obstruction for the existence of $m_n\colon X \times X \to Ht_n X$, $h_{n,n-1} m_n = m_{n-1}$, $h_n F \sim m_n i$, is the class $[\hat{m}_{n-1}] \in [X \wedge X,\ K(\pi_n X,\ n+1)]$. Now, k_n and consequently $k_n m_{n-1}$ has an order power of p and as $H^*(\Lambda,\ \pi_n X)$ is monic $\hat{m}_{n-1}$ too has order power of p. But $H^*(X \wedge X,\ Z)$ is p-torsion free, hence $\hat{m}_{n-1} \sim *$ and m_n exists. $\lim\limits_{\leftarrow} m_n\colon X \times X \to X$ is a mod p H-structure.

Suppose inductively $X \approx_p \hat{X} \times \prod_i B(n_i,\ p)$ and $\hat{X}$ is m-1 connected. Then $\hat{X}$ satisfies the hypothesis of 5.2.3 as well. $\hat{X}_m \to Ht_m \hat{X} = K(\pi_m \hat{X},\ m)$ lifts to $F_{(p)}(K(\pi_m \hat{X},\ m)) = B(m,p)^{\mathrm{rank}(\pi_m \hat{X})}$. On the other hand $B^{k_m} = B(m,p)^{\mathrm{rank}(\pi_m \hat{X})} \to K(\pi_m \hat{X},\ m)$ lifts to $\hat{X}$ to obtain

$$\begin{array}{ccccc}
B^{k_m} & \xrightarrow{g_1} & \hat{X} & \xrightarrow{g_2} & B^{k_m} \\
& \searrow & \downarrow & \swarrow & \\
& & K(\pi_m \hat{X}) & &
\end{array}$$

and by 5.1.2 (b) g_2g_1 is mod p equivalence. It follows that $\hat{X} \approx_p B^{k_n} \times \text{fiber } g_2$, $X \approx \text{fiber } g_2 \times \prod_i B(n_i,p) \times B(m,p)^{\text{rank } \pi_m(\hat{X})}$ and fiber g_2 is at least m connected. The p-equivalence $X \approx_p \Pi B(n_i,p)$ follows from a simple limit process.

In [Wilson] it was actually shown that B(n,p) are (of the mod p type of) ∞-loop spaces, hence, an H_0 space with p-torsion free homotopy and homology is mod p equivalent to an ∞-loop space.

It was also shown by Wilson that $H^*(B(n,p), Z_p)$ are free algebras on generators of dim ≡ n mod 2p-2.

5.2.4. Examples: The 5.2.2 characterization of B(n,p) easily yields: B(2,p) = K(Z,2) for every p. B(3,2) = SU, B(4,2) = BSU. B(3,3) = Sp, B(4,3) = BSp. In general it was proved by Peterson (see [Peterson]) that $SU \approx_p \prod_{i=1}^{p-1} X_i(p)$, $X_i(p)$ mod p indecomposable and 2i connected. For $1 \leq i \leq p-1$ $B(2i+1, p) = X_i(p)$. Similarly $BSU \sim_p \prod_{i=1}^{p-1} Y_i(p)$, $Y_i(p)$ 2i+1 connected and $B(2i+2, p) = Y_i(p)$ $1 \leq i \leq p-1$.

5.3. The groups [, B(n,p)].

5.3.1. Lemma: *Let* M, Y *be spaces*, $H_*(M,Z)$ *and* $\pi_*(Y)$ p-*torsion free*. Y - *an* H_0 *space. Let* $[M,Y]_p$ *be the set of* p-*homotopy classes of maps* (M → Y) *(see 4.2). Then* $H_*(\ ,Q): [M,Y]_p \to \text{Hom}(H_*(M,Q), H_*(Y,Q))$ *is* 1-1.

Proof: We replace $\hat{M}$ by a p-equivalent space M with $H_*(\hat{M},Z)$-torsion free: Inductively construct

$$\begin{array}{ccccccccc} \lim_{\leftarrow} M_n = \hat{M} & \longrightarrow \dots \longrightarrow & M_n & \xrightarrow{h_n} & M_{n-1} & \longrightarrow \dots \longrightarrow & M_0 = M \\ & & & & \downarrow{\scriptstyle k_n} & & \\ & & & & K\,(\text{torsion } H_n(M_{n-1}, Z), n) & & \end{array}$$

h_n - the fibration induced by k_n from the path space fibration. h_n are p-equivalences and so is $h\colon \hat{M} \to M$. $H_m(M_n, Z)$ is torsion free for $m \leq n$ and consequently $H_*(\hat{M},Z)$ is torsion free. Similarly, replace Y by a p-equivalent space $\hat{Y}$ with $\pi_*(\hat{Y})$ torsion free: Let $\psi\colon Y \xrightarrow{\sim 0} K(\pi_*(Y)/\text{torsion})$. Put $\hat{Y} = Y(\{p\}, \psi)$ in the terminology of 4.3, $\psi'\colon Y \xrightarrow{\sim p} \hat{Y}$. We shall show that $[\hat{M},\hat{Y}] \xrightarrow{H_*(\ ,Q)} \mathrm{Hom}(H_*(\hat{M},Q), H(\hat{Y},Q))$ is injective and then considering

$$\begin{array}{ccc} [M,Y] & \xrightarrow{[h,\ \psi']} & [\hat{M},\hat{Y}] \\ \downarrow & & \downarrow \\ \mathrm{Hom}(H_*(M,Q), H_*(Y,Q)) & \xrightarrow[\approx]{} & \mathrm{Hom}[H_*(\hat{M},Q), H_*(\hat{Y},Q)] \end{array}$$

$H_*(f,Q) = H_*(g,Q)$ implies $\psi' fh \sim \psi' gh$ and $f \sim_p g$, which will show that $[M,Y]_p \to \mathrm{Hom}[H_*(M,Q), H_*(Y,Q)]$ is injective (and moreover, $[M,Y]_p = [\hat{M},\hat{Y}]$). To show that $[\hat{M},\hat{Y}] \to \mathrm{Hom}(H_*(\hat{M},Q), H_*(\hat{Y},Q))$ is injective we shall show that if $H_*(f_1, Q) = H_*(f_2, Q)$ then any homotopy $Ht_{n-1}(f_1) \sim Ht_{n-1}(f_2)\colon \hat{M} \to Ht_{n-1}(\hat{Y})$ can be (essentially) lifted to a homotopy $Ht_n(f_1) \sim Ht_n(f_2)$ and one can pass to a limit $f_1 \sim f_2$. Indeed, a homotopy $Ht_{n-1}(f_1) \sim Ht_{n-1}(f_2)$ induces a map $w\colon \hat{M} \to K(\pi_n(\hat{Y}),n)$ so that $w * Ht_n(f_1) \sim Ht_n(f_2)$. This homotopy covers the homotopy

$Ht_{n-1}(f_1) = h_{n,n-1}(w * Ht_n(f_1)) \sim h_{n,n-1}Ht_n(f_2) = Ht_{n-1}(f_2)$

$(h_{n,n-1}: Ht_n(\hat{Y}) \to Ht_{n-1}(\hat{Y}))$.

As $\hat{Y}$ is an H_0 space $\hat{Y} \approx_0 K(\pi_*(Y))$ and there exists $u: Ht_n(\hat{Y}) \to K(\pi_n(\hat{Y}), n)$ so that the composition

$$K(\pi_n(\hat{Y}), n) \xrightarrow{j} Ht_n(\hat{Y}) \xrightarrow{u} K(\pi_n(\hat{Y}), n)$$

is a 0-equivalence. One can follow u by an 0-equivalence if necessary so that one may assume $uj = \lambda \cdot 1$, $0 \neq \lambda \in Z$.

If $\eta: K(\pi_n(Y), n) \times Ht_n(Y) \to Ht_n(Y)$ is the principal action $(w * Ht_n(f_1) = \eta(w \times f_1)\Delta_{\hat{M}})$ one has $u\eta = \lambda p_1 + up_2$. As $H_*(Ht_n(f_1), Q) = H_*(Ht_n(f_2), Q)$ $H_*(uHt_n(f_1), Q) = H_*(uHt_n(f_2), Q)$. But as $H_*(\hat{M},Q)$, $\pi_*(\hat{Y})$ are torsion free

$$[\hat{M}, K(\pi_n\hat{Y}, n)] \to Hom(H_n(\hat{M},Z), H_n(K(\pi_n\hat{Y}, n), Z) = \pi_n\hat{Y}) \to$$
$$\to Hom(H_n(\hat{M},Q), H_n(K(\pi_n\hat{Y}, n), Q) = \pi_n\hat{Y} \otimes Q)$$

are injective and $uHt_n(f_1) \sim uHt_n(f_2)$.

$$uHt_n(f_2) = u\eta(w \times Ht_n(f_1))\Delta_{\hat{M}} = \lambda w + uHt_n(f_1).$$

Hence $0 = \lambda w \in H^n(\hat{M}, \pi_n\hat{Y})$. As the latter is torsion free $w \sim *$ and corresponds to a homotopy $w * Ht_n(f_1) \sim Ht_n(f_1)$

corel $h_{n,n-1}$. The homotopy $Ht_n(f_2) \sim w*Ht_n(f_1) \sim Ht_n(f_1)$ covers the homotopy $Ht_{n-1}(f_2) \sim Ht_{n-1}(f_1)$.

5.3.1.1. Remark: The homotopy described in the proof of 5.3.1 $\Phi_n: Ht_n(f_1) \sim Ht_n(f_2)$ technically does not cover $\Phi_{n-1}: Ht_{n-1}(f_1) \sim Ht_{n-1}(f_2)$. It covers the latter followed by a constant homotopy k with non zero length ℓ_n: $Ph_{n,n-1}\tilde{\phi}_n = \phi_{n-1} + k_{\ell_{n-1}}$ where $\ell_{n-1}: Y \to R^+$. However, one can easily choose ϕ_n, ℓ_n to satisfy $\sum_{n=0}^{\infty} \ell_n(y) < \infty$ so that if $\ell^{(n)}(y) = \sum_{m \geq n} \ell_m(y)$ then $\phi_n + k_{\ell^{(n)}}$ covers $\phi_{n-1} + k_{\ell^{(n-1)}}$.

As a direct consequence of 5.1.2 (a) and the definition of B(n,p) one has:

5.3.2. Lemma: *Let* $\sigma: [X, B(n,p)] \to H^n(X, Z_p)$ *be induced by* $B(n,p) \to K(Z_p, n)$ *(yielding an isomorphism of* $H^n(\ , Z_p)$*). If* $H_*(X,Z)$ *is* p *torsion free then* σ *is surjective.*

5.3.3. Proposition: *Let* X *be a space with* $H_*(X,Z)$ p-*torsion free. Then there exists* $\alpha: X \to \prod_i B(n_i, p)$ *so that* $H^*(\alpha, Z_p)$ *is surjective.*

Proof: For any group G $F_{(p)}(K(G,n)) \approx_p B(n,p)^a$ where $a = \text{rank } G \otimes Z_p$. If $H_*(X,Z)$ is p-torsion free $\hat{\alpha}: X \to K(H_*(X), Z)$ lifts to $\alpha: X \to F_{(p)}(KH_*(X,Z)) \approx \prod_n B(n,p)^{a_n}$, $a_n = \text{rank}(H_*(X,Z) \otimes Z_p) = \text{rank } H_*(X,Z_p)$. α is the desired map.

5.3.4. Proposition: *Given spaces* L, M *and an* H_0 *space* Y. *Suppose* $H_*(M,Z)$ *and* $\pi_*(Y)$ *are* p-*torsion free. If* $f: L \to M$ *yields a surjection* $H^*(f,Z)$ *then for some* p-*equivalent* $\hat{Y}$ *of* Y

$$f^*: [M,\hat{Y}] \to [L,\hat{Y}]$$

is surjective.

Proof: As in the proof of 5.3.1 there exists $\hat{Y}$, $\hat{Y} \sim_p Y$ so that $\pi_*(\hat{Y})$ is torsion free and the k-invariants of $\hat{Y}$ are of order power of p.

Given $g: L \to \hat{Y}$. Suppose one has a commutative diagram

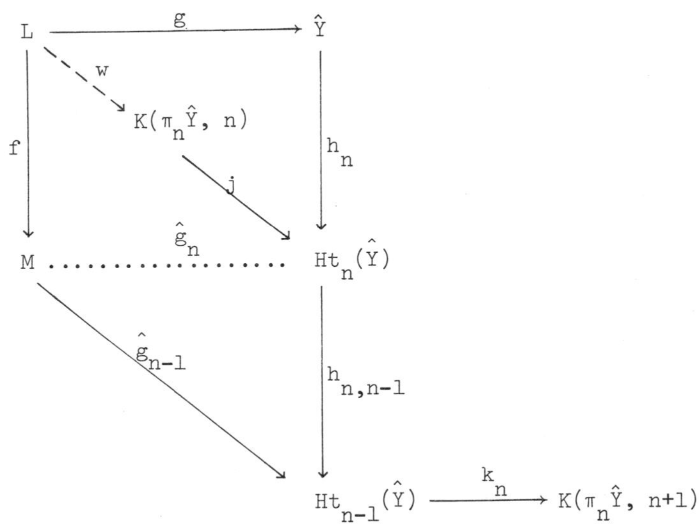

and a homotopy $\phi_{n-1}: L \to PHt_{n-1}(\hat{Y})$ of $h_{n,n-1}h_n g \sim \hat{g}_{n-1}f$. As $\pi_n\hat{Y}$ is free, k_n of order p^{r_n} and $H^{n+1}(M,Z)$ p-torsion free $\hat{g}_{n-1}$ lifts to $\hat{g}'_n$. If one considers $h_{n,n-1}$ as a fibration the homotopy ϕ_{n-1} can be lifted to a homotopy $\phi: L \to PHt_n(\hat{Y})$, $\phi: h_n g \sim w * (\hat{g}'_n f)$ for some $w: L \to K(\pi_n(\hat{Y}), n)$.

As $H^n(f,Z)$ and consequently $H^n(f, \pi_n(\hat{Y}))$ are surjective $w \sim \hat{w}f$, $\hat{w}: M \to K(\pi_n(\hat{Y}), n)$. Replace $\hat{g}'_n$ by $\hat{g}_n = \hat{w} * \hat{g}'_n$ which lifts $\hat{g}_{n-1}$ as well and $\phi: h_n g \sim w * (\hat{g}'_n f)$ can be followed by a homotopy corel $h_{n,n-1}$ $w * (\hat{g}'_n f) \sim \hat{w}f * (\hat{g}'_n f) = g_n f$ to obtain $\phi_n: h_n g \sim \hat{g}_n f$ lifting ϕ_{n-1} (and see remark 5.3.1.1). One can pass to a limit $\hat{g} = \lim\limits_{\leftarrow} \hat{g}_n: M \to \hat{Y}$, $\phi = \lim\limits_{\leftarrow} \phi_n: g \sim \hat{g}f$.

Putting together 5.3.1, 5.3.2, 5.3.3 and 5.3.4 and choosing $B(n,p)$ to have p-torsion integral k-invariants one obtains:

5.3.5. Corollary: *If* $H_*(X,Z)$ *is p-torsion free then*

(a) $\sigma: [X, B(n,p)] \to H^n(X, Z_p)$ *is surjective*

(b) $[X, B(n,p)]_p \to \mathrm{Hom}(H_*(X,Q), H_*(B(n,p), Q))$ *is injective*

(c) If $f: X' \to X$ *yields a surjection* $H^*(f,Z)$ *then* $f^*: [X, B(n,p)] \to [X', B(n,p)]$ *is a surjection.*

(d) There exists $\alpha: X \to \prod_i B(n_i, p)$ *so that* $H^*(\alpha, Z_p)$ *is surjective.*

5.4. H-maps into B(n,p)

We study here conditions for H-maps $X \to K(Z_p, n)$ to be H-liftable to $X \to B(n,p)$. We choose $B(n,p)$ to have integral p-torsion k-invariants.

5.4.1. Proposition: *Let* X, μ *be an H-space,* $H_*(X,Z)$ *torsion free. Given a map* $\alpha: X \to \prod_i B(n_i, p) = B(X)$ *so that* $H^*(\alpha,Z)$ *is surjective. There exists an H-structure for* $B(X)$ *so that* α *is an H-map.*

Proof: $H^*(\alpha,Z)$ surjective implies

$$H^*(\alpha \wedge \alpha, Z): H^*(B(X) \wedge B(X), Z) \to H^*(X \wedge X, Z)$$

surjective (as $H^*(X,Z)$ is torsion free). By 5.3.4

$(\alpha \wedge \alpha)^*$: $[B(X) \wedge B(X), B(X)] \to [X \wedge X, B(X)]$ is surjective and $HD(\alpha, \mu, m_{B(X)}) \in \mathrm{im}(\alpha \wedge \alpha)^*$. Now apply 1.5.1 (a).

<u>5.4.2. Proposition</u>: *Let* X, μ *be an* H-*space. Suppose* $H^*(X,Z)$ *and coker* $\bar{\mu}^*$ *are* p-*torsion free.*

$$(\bar{\mu}^* = H^*(\mu,Z) - H^*(p_1, Z) - H^*(p_2,Z) \colon H^*(X,Z) \to H^*(X \wedge X, Z))$$

Then for any primitive element $x \in H^n(X, Z_p)$ *there exists an* H-*map* $f_x \colon X \to B(n,p)$ *so that* $\sigma[f_x] = x$.

5.4.2 follows from the more general proposition.

<u>5.4.3. Proposition</u>: *Let* X, μ *be as in* 5.4.2. *Given* H-*spaces and an* H-*map* $h \colon \tilde{Y}, \tilde{\mu} \to Y, \mu_0$ *so that* $\pi_*(\text{fiber } h)$ *are free and suppose there exists a commutative diagram of* H-*spaces and maps:*

$$\begin{array}{ccc} \tilde{Y}, \tilde{\mu} & \xrightarrow[\sim \mathbb{P} - p]{\alpha} & Y \times K(\pi_*(\text{fiber } h)), \hat{\mu} \\ & {\scriptstyle h} \searrow \quad \swarrow {\scriptstyle p_1} & \\ & Y, \mu_0 & \end{array}$$

$\hat{\mu}$ - *the product multiplication. Then any* H-*map* $f \colon X, \mu \to Y, \mu_0$ *lifts to an* H-*map* $\tilde{f} \colon X, \mu \to \tilde{Y}, \tilde{\mu}$.

<u>Proof</u>: One uses a Postnikov decomposition of h and induction:

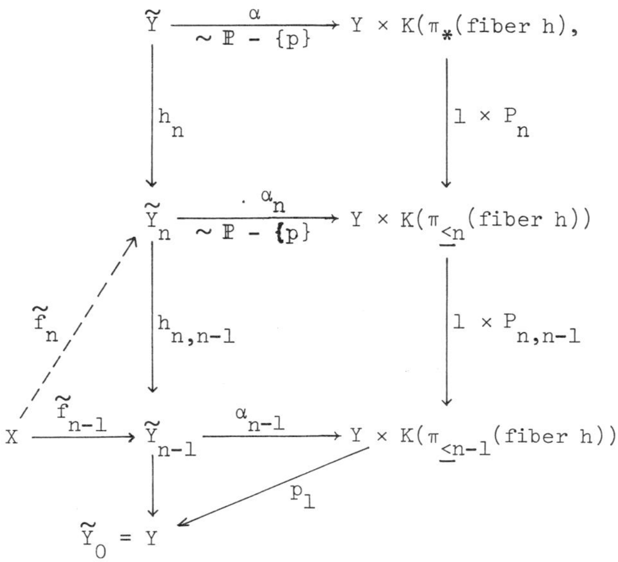

Given an H-map $\tilde{f}_{n-1}, \tilde{F}_{n-1}\colon X, \mu \to \tilde{Y}_{n-1}, \tilde{\mu}_{n-1}$, the lifting of $\tilde{f}_{n-1}$ to a map $\tilde{f}'_n\colon X \to \tilde{Y}_n$ exists by our standard k-invariant argument.

$HD(\tilde{f}_n, \mu, \tilde{\mu}_n) = jw_n$,

$$w_n\colon X \wedge X \longrightarrow K(\pi_n(\text{fiber } h),\ n)$$

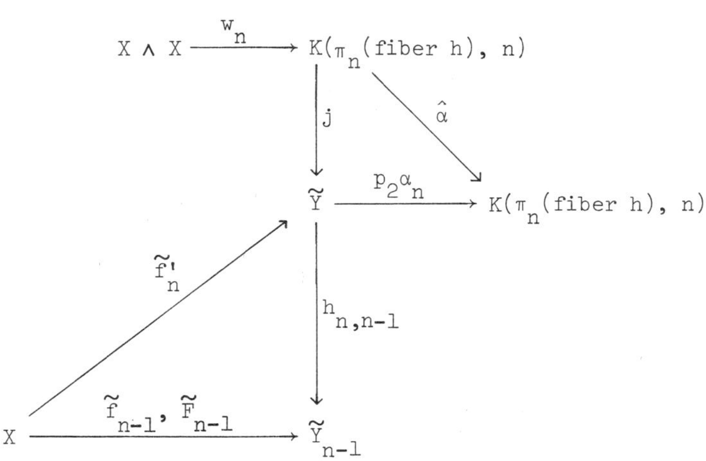

The homotopy $\widetilde{F}'_n\colon jw_n + \widetilde{\mu}_n(\widetilde{f}'_n \times \widetilde{f}'_n) \sim \widetilde{f}'_n\mu$ covers $\widetilde{F}_{n-1}$, $Ph_{n,n-1}\widetilde{F}'_n \sim \widetilde{F}_{n-1}$ (corel E_0, E_∞). (We assume $h_{n,n-1}$ to be a fibration and liftings into $\widetilde{Y}_n$ to be exact liftings not only up to homotopy).

As $p_2\alpha_n$ is an H-map

$$HD(p_2\alpha_n\widetilde{f}'_n,\ \mu,\ m_K) = p_2\alpha_n HD(\widetilde{f}'_n,\ \mu,\ \widetilde{\mu}_n) = p_2\alpha_n jw_n = \hat{\alpha}w_n.$$

Now, $HD(p_2\alpha_n\widetilde{f}'_n,\ \mu,\ m_k) \in \operatorname{im}(\overline{\mu}^* \otimes 1)$

$$\overline{\mu}^* \otimes 1\colon\ H^n(X,\ \pi_n(\text{fiber } h)) = H^n(X,Z) \otimes \pi_n(\text{fiber } h) \to$$
$$\to H^n(X \wedge X,\ Z) \otimes \pi_n(\text{fiber } h) = H^n(X \wedge X,\ \pi_n(\text{fiber } h)).$$

As coker $\overline{\mu}^*$ (hence, coker $\overline{\mu}^* \otimes 1$) is p-torsion free and $H^*(\hat{\alpha},Z)$ $\mathbb{P}$-p isomorphism $\hat{\alpha}w_n \in \operatorname{im}(\overline{\mu}^* \otimes 1)$ implies $w_n \in \operatorname{im} \overline{\mu}^* \otimes 1$, $w_n = (\overline{\mu}^* \otimes 1)v$, $v\colon X \to K(\pi_n(\text{fiber } h),\ n)$. Alter $\widetilde{f}'_n$ to a lifting $\widetilde{f}_n$, $\widetilde{f}'_n = jv + \widetilde{f}_n$ and essentially by 2.2.1 $HD(\widetilde{f}_n,\ \mu,\ \widetilde{\mu}_n) = *$ and furthermore, one has an H-structure $\widetilde{F}_n$ for $\widetilde{f}_n$ so that $Ph_{n,n-1}\widetilde{F}_n \sim \widetilde{F}_{n-1}$ (corel E_0, E_∞). This observation allows us to pass to a limit $\varprojlim(\widetilde{f}_n,\ \widetilde{F}_n) = \widetilde{f},\ \widetilde{F}\colon X \to \varprojlim \widetilde{Y}_n = \widetilde{Y}$.

To apply 5.4.3 for the proof of 5.4.2 take $Y = K(Z_p,\ n)$, $\widetilde{Y} = B(n,p)$. The existence of α follows from the observatoon that $H^*(B(n,p),\ Q)$ is primitively generated and one has a 0-equivalence and an H-map $\psi\colon B(n,p) \to K(\pi_*(B(n,p)))$. Replace $B(n,p)$ by $B(n,p)(\{p\},\ \psi)$ if necessary to obtain $\psi''\colon B(n,p) \xrightarrow[\sim\ \mathbb{P}\text{-p}]{H} K(\pi_*(B(n,p)))$ and

$$\begin{array}{ccc} B(n,p) & \xrightarrow{(\sigma\ \times\ \psi'')\Delta} & K(Z_p,n)\ \times\ K(\pi_*(B(n,p))) \\ & \searrow^{\sigma} \qquad \swarrow_{p_1} & \\ & K(Z_p,n) & \end{array}$$

The condition on coker $\overline{\mu}^*$ in 5.4.2 and 5.4.3 can be recognized using the following:

<u>5.4.4. Lemma</u>: *Let* A, , ψ *be a torsion free connected graded Hopf algebra of finite type over* Z. *Suppose* A *is cocommutative and coassociative. Then* coker $(\overline{\psi}: \overline{A} \to \overline{A} \otimes \overline{A})$ *is* p-*torsion free if and only if the dual of* $A \otimes Z_p$ *is a free algebra.*

<u>Proof</u>: As ψ is coassociative $(\overline{\psi} \otimes 1)\overline{\psi} = (1 \otimes \overline{\psi})\overline{\psi}$ and $\text{im}\ \overline{\psi} \subset \ker(\overline{\psi} \otimes 1 - 1 \otimes \overline{\psi})$. Cocommutativity implies $\text{im}\ \overline{\psi} \subset \ker(1-T)$. As $(1-T)\ker(\overline{\psi} \otimes 1 - 1 \otimes \overline{\psi}) \subset \ker(\overline{\psi} \otimes 1 - 1 \otimes \overline{\psi})$, $\ker(\overline{\psi} \otimes 1 - 1 \otimes \overline{\psi}) = \ker(1-T) \cap \ker(\overline{\psi} \otimes 1 - 1 \otimes \overline{\psi}) \oplus W$. $\overline{A} \otimes \overline{A} = \ker(\overline{\psi} \otimes 1 - 1 \otimes \overline{\psi}) \oplus V$ $(V = \text{im}(\overline{\psi} \otimes 1 - 1 \otimes \overline{\psi}) \subset \overline{A} \otimes \overline{A} \otimes \overline{A}$ which is free).

$\overline{A} \otimes \overline{A} = [\ker(1-T) \cap \ker(\overline{\psi} \otimes 1 - 1 \otimes \overline{\psi})] \oplus W \oplus V$ and $\text{im}\ \overline{\psi}$ lies in the first summand.

Coker $\overline{\psi}$ is p-torsion free if and only if

$$U = [\ker(1-T) \cap \ker(\overline{\psi} \otimes 1 - 1 \otimes \overline{\psi})]/\text{im}\ \overline{\psi}$$

is p-torsion free. Now $\ker(\overline{\psi} \otimes 1 - 1 \otimes \overline{\psi})/\text{im}\ \overline{\psi} \approx \text{Cotor}_A^{2,*}(Z,Z)$ and for every field F if $A_F, \phi_F, \psi_F = A \otimes F,\ \phi \otimes F,\ \psi \otimes F$ then $\ker(\overline{\psi}_F \otimes 1 - 1 \otimes \overline{\psi}_F)/\text{im}\ \overline{\psi}_F = \text{Cotor}_{A_F}^{2,*}(F,F)$ and

$$U \otimes F = [\ker(1-T) \cap \ker(\overline{\psi}_F \otimes 1 - 1 \times \overline{\psi}_F)]/\text{im}\ \overline{\psi}_F = Q\ \text{Cotor}_{A_F}^{2,*}(F,F).$$

For $F = Q$ $Q\ \text{Cotor}^{*,*}(F,F) \subset Q\ \text{Cotor}_{A_F}^{1,*}(F,F)$ and $U \otimes Q = 0$. For $F = Z_p$ $Q\ \text{Cotor}_{A_F}^{2,*}(F,F) = 0$ if and only if $(A_F)^*$ is a free algebra. $U \otimes Q = 0$ implies that U is torsion and $U \otimes Z_p = 0$ if and only if coker $\overline{\psi}$ is p-torsion free.

As a consequence of 5.4.2 and 5.4.4 one obtains:

5.4.5. Corollary: *Let* X, μ *be an* H-*space* $H^*(X,Z)$ p-*torsion free* $H^*(X,Q)$ *primitively generated and* $H_*(X, Z_p)$ *a free (associative commutative graded) algebra. Then every* $x \in PH^n(X, Z_p)$ *can be realized by an* H-*map* $\hat{f}: X \to B(n,p)$ $\sigma[\hat{f}] = x$.

Proof: $A = H^*(X,Z)/\text{torsion}$ is a Hopf algebra and as $A \otimes Q$ is primitively generated $A \otimes Q$ is coassociative and cocommutative and so is A. As $H^*(X,Z_p) = A \otimes Z_p$ by 5.4.4 if

$$\bar{\mu}^*: \bar{H}^*(X,Z)/\text{torsion} \to \bar{H}^*(X,Z)/\text{torsion} \otimes \bar{H}^*(X,Z)/\text{torsion}$$

then coker $\bar{\mu}^*$ is p-torsion free, and as

$$\tilde{H}^*(X,Z) \xrightarrow{\approx p} \tilde{H}^*(X,Z)/\text{torsion}$$

$$\tilde{H}^*(X \wedge X, Z) \xrightarrow{\approx p} \tilde{H}^*(X,Z)/\text{torsion} \otimes \tilde{H}^*(X,Z)/\text{torsion}$$

Coker $\bar{H}^*(\mu,Z)$ is p-torsion free and one can apply 5.4.2.

5.4.5.1. Remark: The non redundancy of the condition on the algebra $H_*(X,Z_p)$ in 5.4.5 can be seen by the following simple example: $x_2^p: K(Z,2) \to K(Z_p, 2p)$. This H-map cannot be lifted to an H-map $f: K(Z,2) \to B(2p,p)$ as the latter is a mod p H-retract of BSU, hence $H_*(B(2p,p), Z_p)$ is a free algebra while in $H_*(K(Z,2),Z_p)$ pth powers vanish.

5.5. Examples: Some properties of BU

It is classicaly known that $BU(1) = CP^{\infty} \to BU$ induces an isomorphism $PH^*(BU, Z_p) \xrightarrow{\approx} H^*(CP^{\infty}, Z_p)$, hence, $PH^{2n}(BU, Z) = Z$, $PH^{2n+1}(BU,Z) = 0$ and if $0 \neq v_{2n} \in PH^{2n}(BU, Z_p)$ then $P^i v_{2n} = c\binom{n}{i} v_{2n+2i(p-1)}$, $0 \neq c \in Z_p$. As a consequence one can easily see that

5.5.1.1. $PH^*(BU, Z_p)$ is generated over $A(p)$ by $\bigoplus_{\substack{a<p \\ 0\leq j}} PH^{2ap^j-2}(BU, Z_p)$.

5.5.1.2. $P^{p-1} v_{2ap-2} = c v^p_{2(p+a-2)}$ $0 \neq c \in Z_p$.

One has the following:

5.5.2. Lemma: *Let* X, μ *be an H-space,* $f\colon X \to BU$ *an H-map. If* $H^*(X, Z_p)$ *is a free algebra and if* $H^{2n}(f, Z_p)$ *is injective for some* $n < p$ *then* $PH^{2n+2k(p-1)}(f, Z_p)$ *is injective for all* k.

Proof: We prove for p-odd. $p = 2$ is similar. Suppose $PH^{2m}(f,Z_p)$ is injective, for $m = k(p-1)+n$, $k < k_0$. Let $m_0 = k_0(p-1)+n = m_0'p + m_0''$, $0 \leq m_0'' < p$. Then $P^{m_0''} v_{2m_0} = c v^p_{2(m_0'+m_0'')}$, $0 \neq c \in Z_p$. $m_0' + m_0'' \equiv n$ mod(p-1). As $H^*(X,Z_p)$ is a free algebra $H^*(f,Z_p)v_{2(m_0'+m_0'')} \neq 0$ implies $0 \neq H^*(f,Z_p)v^p_{2(m_0'+m_0'')} = P^{m_0''} H^*(f,Z_p)v_{2m_0}$ and $H^{2m_0}(f,Z_p)v_{2m_0} \neq 0$. (as $H^*(X,Z_p)$ is a free algebra) and $PH^{2m_0}(f, Z_p)v_{2m_0} \neq 0$.

5.5.3. Proposition ([Peterson]): $BSU \approx_p \prod_{n=2}^{p} B(2n,p)$.

Proof: Let $\psi_0\colon BSU \xrightarrow{\approx 0} \prod_{n=2}^{\infty} K(Z, 2n)$. Put $\hat{BSU} = BSU(\{p\}, \psi)$. Hence the k-invariants of $\hat{BSU}$ are integral of order power of p.

$Ht_{2p}(\hat{BSU}) = \prod_{n=2}^{p} K(Z,2n)$. Lift $\hat{BSU} \to \prod_{n=2}^{p} K(Z, 2n)$ to $\alpha_2 \colon \hat{BSU} \to \prod_{n=2}^{p} B(2n,p)$ and lift $\prod_{n=2}^{p} B(2n,p) \to Ht_{2p}(\hat{BSU}) = \prod_{n=2}^{p} K(Z,2n)$ to $\alpha_1 \colon \prod_{n=2}^{p} B(2n,p) \to \hat{BSU}$.

$$\begin{array}{ccccc} \prod_{n=2}^{p} B(2n,p) & \xrightarrow{\alpha_1} & \hat{BSU} & \xrightarrow{\alpha_2} & \prod_{n=2}^{p} B(2n,p) \\ & \searrow & \downarrow & \swarrow & \\ & & \prod_{n=2}^{p} K(Z, 2n) & & \end{array}$$

by 5.1.2 (b) $\alpha_2\alpha_1$ is a p-equivalence.

Looping twice one obtains

$$\Omega^2\Big(\prod_{n=2}^{p} B(2n,p)\Big) \xrightarrow{\Omega^2\alpha_1} \hat{BU} = \Omega^2\hat{BSU} \xrightarrow{\Omega^2\alpha_2} \Omega^2\Big(\prod_{n=2}^{p} B(2n,p)\Big).$$

As $\Omega^2\alpha_1$ is a homotopy equivalence through dim $2p$ $PH^{2n}(\Omega^2\alpha_1, Z_p)$ injective for $n < p$. By 5.5.2 (as only mod p arguments are used it applies to $\hat{BU}$ as well) $PH^*(\Omega^2\alpha_1, Z_p)$ and consequently $H^*(\Omega^2\alpha_1, Z_p)$ are injective. $\Omega^2\alpha_2\Omega^2\alpha_1$ p-equivalence implies $H^*(\Omega^2\alpha_1, Z_p)$ surjective, hence $\Omega^2\alpha_1$, consequently α_1 and α_2 are p-equivalences and so is the composition $BSU \to \hat{BSU} \xrightarrow{\alpha} \prod_{n=2}^{p} B(2n,p)$. As $BU \approx K(Z,2) \times BSU$, $BU \approx_p \prod_{n=1}^{p} B(2n,p)$ $(B(2,p) \approx K(Z,2))$. Similarly one can argue that $SU \approx_p \prod_{n=1}^{p-1} B(2n+1, p)$.

<u>5.5.4. Corollary</u>: *Let* X, μ *be an H-space.* $H^*(X,Q)$ *primitively generated,* $H_*(X,Z)$ p-*torsion free and* $H^*(X,Z_p)$ *and* $H_*(X,Z_p)$ *free algebras. If* $x \in PH^{2n}(X, Z_p)$ $n < p$ *then there exists an H-map*

$f: X \to B(2n,p)$, $x \in \operatorname{im} H^*(f, Z_p)$ *and* $H^*(f, Z_p)$ *injective.*

Proof: By 5.4.5 an H-map f exists. By $BU \approx_p \prod_{n=1}^{p} B(2n,p)$ and 5.5.2 $PH^*(f, Z_p)$ and consequently $H^*(f, Z_p)$ are injective.

5.5.5. Corollary: *Let* X, μ *be an* H-*space. Then* $H^*(X,F) \approx H^*(BU, F)$ $F = Z_p$ *and* Q *as Hopf algebras if and only if* X, μ *is* H-p *equivalent to* $\hat{B}U = BU(\{p\}, \psi)$, $(\psi: BU \xrightarrow{\approx 0} \prod_{n=1}^{\infty} K(Z, 2n))$.

Proof: One can obtain an H-map

$$f_{2p}: X \to Ht_{2p}(\hat{B}U) \approx \prod_{n=1}^{p} K(Z, 2n).$$

By 5.4.5 f_{2p} lifts to an H-map $f: X \to BU$. By 5.5.2 $H^*(f, Z_p)$ is injective and as $H^*(X,Z_p) \approx H^*(BU, Z_p)$ f is a p-equivalence.

5.5.6. Corollary: *Let*

$$G_H(BU) = \{X, \mu \mid \text{for every } p \text{ there exists } f_p: X \xrightarrow[H]{\sim p} BU\}.$$

Then $X, \mu \in G_H(BU)$ *if and only if* $H^*(X,Z) \approx H^*(BU,Z)$ *as a Hopf algebra. Moreover, in this case for every finite set of primes* $\mathbb{P}_1$ *there exists an* H-*map* $f: X \xrightarrow{\approx \mathbb{P}_1} BU$.

Proof: If $X \in G_H(BU)$ $H^*(X,Z_p) \approx H^*(BU,Z_p)$ as a Hopf algebra for every p. This implies $H^*(X,Z) \approx H^*(BU,Z)$. If $H^*(X,Z) \approx H^*(BU,Z)$ then $H^*(X,F) \approx H^*(BU,F)$ for $F = Z_p$ or Q. By 5.5.5 for every p there exists an H p-equivalence $f_p: X \xrightarrow[H]{\sim p} BU$ and $X, \mu \in G_H(BU)$.

If $\mathbb{P}_1$ is a finite set of primes, as BU is HA and HC

$$f_{\mathbb{P}_1} = \sum_{p \in \mathbb{P}_1} \frac{a}{p} f_p : X \to BU \quad (a = \prod_{p \in \mathbb{P}_1} p)$$ is an H-$\mathbb{P}_1$ equivalence.

One can easily argue (and see 4.6.8) that $G_H(BU)$ is uncountable.

<u>5.5.7. Corollary</u>: *Let* X *be a CW complex,* $H^*(X, Z_p)$ *an exterior algebra on odd dimensional generators. If*

$$\text{rank } QH^{2n+2k(p-1)+1}(X, Z_p) = a_n$$

is independent of k *and if* $P^{2m}: QH^{2m+1}(X,Z_p) \to QH^{2mp+1}(X,Z_p)$ injective then $X \approx_p \prod_{n=1}^{p-1} B(2n+1, p)^{a_n}$.

<u>Proof</u>: $H^*(\Omega X, Z_p)$ and $H_*(\Omega X, Z_p)$ are free algebras, $H^*(X, Q)$ primitively generated and one can apply 5.5.4.

5.6. Non stable BP Adams resolutions.

<u>5.6.1. Definition</u>: Given an Ω-spectrum $E_* = \{E_n, \gamma_n\}$ and a space X, *a strong* E_* *Adams resolution for* X is a sequence of spaces and maps

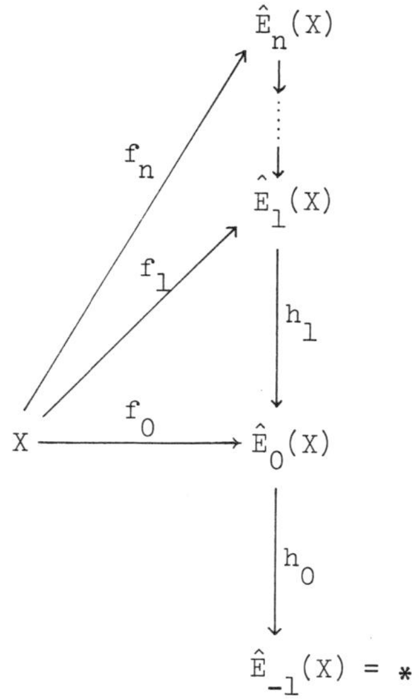

so that

(a) $h_n: \hat{E}_n(X) \to \hat{E}_{n-1}(X)$ is a principal fibration induced by a map

$$k_n: \hat{E}_{n-1}(X) \to \prod_{j \in J_n} E_{m_j}$$

(b) If the fiber of f_n is a_n connected then $\lim\limits_{n\to\infty} a_n = \infty$.

(c) For every i and n, $f_0^*: [\hat{E}_0(X), E_i] \to [X, E_i]$ is surjective and

$$\ker\{[\hat{E}_n(X), E_i] \to [\hat{E}_{n+1}(X), E_i]\} = \ker\{[\hat{E}_n(X), E_i] \to [X, E_i]\}.$$

<u>5.6.1.1. Remark</u>: Note that the classical $K(Z_p, \ )$ Adams resolution for spaces X with $H_*(X,Q) \neq 0$ does not satisfy 5.6.1 (b) but some weaker property holds to obtain an Adams spectral sequence, hence, the resolution is still a good estimate for X. The efficiency of E_* as the underlying spectrum for an Adams resolution for X is measured by the rate of

increase of a_n in 5.6.1 (b). e.g.: For p-torsion spaces (i.e.: spaces with $H_*(\ ,Z)$ being p-torsion) the $K(\ , Z_p)$ resolution yields a monotonic non decreasing sequence $\{a_n\} \to \infty$.

$\{B(n,p)\}$ do not form an Ω-spectrum as $B(n,p) \approx \Omega B(n+1, p)$ only for $n \neq 2 \frac{p^r - 1}{p - 1} = v(r)$ while $\Omega B(v(r) + 1, p) \approx B(v(r), p) \times B(p \cdot v(r), p)$. But these relations are enough to establish Adams resolutions using $\{B(n,p)\}$. We shall refer to such a resolution as *a (non stable) BP Adams resolution.*

We shall outline the construction of the resolution for spaces X with $H_*(X,Z)$ p-torsion free and $H^*(X, Z_p)$ a free algebra. A similar construction can be obtained for an H-space X with $H_*(X,Z)$ p-torsion free.

5.6.2. Proposition: *Let* $f: X_1 \to X_2$. *Suppose* $H_*(X_i, Z)$ *and* $H^*(X_1, Z)//\text{im } f^*$ *are* p-*torsion free,* $H^*(X_i, Z_p)$ *and* $H^*(X_1, Z_p)//\text{im } f^*$ *are free algebras. Let* $Y = \text{fiber } f$. *Then:* $H_*(Y,Z)$ *is* p-*torsion free and the Eilenberg-Moore spectral sequence:*

$$\text{Tor}^{H^*(X_2, F)}(H^*(X_1, F), F) \Longrightarrow H^*(Y,F)$$

collapses for $F = Z_p$ *and* Q. *Moreover, if* M_2 *is the free algebra generated by* $\ker QH^*(f,F)$ *then for* $F = Z_p$ *or* Q

$$H^*(Y,F) \approx [H^*(X_1, F)//\text{im } H^*(f,F)] \otimes \text{Tor}^{M_2}(F,F)$$

as an $H^*(X_1, F)$ *module.*

Proof: One can easily argue that under the above hypothesis $QH^*(X_i, Z)$ are p-torsion free and if A_i is the free Z-algebra generated by $QH^*(X_i, Z)$/torsion then $A_i \otimes F \approx H^*(X_i, F)$ $F = Z_p$ and Q. Furthermore, f induces a homomorphism $A(f): A_2 \to A_1$ so that $A(f) \otimes F = H^*(f,F)$. (Actually one can replace X_i, f by p-equivalents so that $A_i = H^*(X_i, Z)$/torsion. This is done by starting with a commutative diagram

$$\begin{array}{ccc} X_1 & \xrightarrow{\quad f \quad} & X_2 \\ \psi_1 \Big\downarrow \approx 0 & & \psi_2 \Big\downarrow \approx 0 \\ K(X_1) = K(\pi_*(X_1)/\text{torsion}) & \xrightarrow{\quad f_0 \quad} & K(\pi_*(X_2)/\text{torsion}) = K(X_2) \end{array}$$

then use 4.3.1 to obtain

$$\begin{array}{ccccc} X_1 & \xrightarrow{\psi_1'} & X_1(\{p\}, \psi_1) & \longrightarrow & K(X_1) \\ \Big\downarrow f & & \Big\downarrow \hat{f} = f(\{p\}, \psi_1, \psi_2) & & \Big\downarrow f_0 \\ X_2 & \underset{\sim p}{\xrightarrow{\psi_2'}} & X_2(\{p\}, \psi_2) & \longrightarrow & K(X_2) \end{array}$$

$\hat{X}_i = X_i(\{p\}, \psi_i)$ are the desired spaces). Now, one has $A_2 = M_1 \otimes M_2$, $A_1 = N_1 \otimes N_2$, M_i, N_i free algebras, M_2 generated by $\ker QA(f)$, N_1 by $\operatorname{im} QA(f)$, $N_1 \approx M_1$. Then $\operatorname{Tor}^{A_2 \otimes F}(A_1 \otimes F, F) \approx (N_2 \otimes F) \otimes \operatorname{Tor}^{M_2 \otimes F}(F,F)$. As $\operatorname{rank}(M_i \otimes F)$, $\operatorname{rank}(N_i \otimes F)$ are independent of F ($F = Z_p$ or Q) the same holds for $\operatorname{Tor}^{A_2 \otimes F}(A_1 \otimes F, F)$.

It suffices to show that the Eilenberg-Moore spectral sequence ${}^{F}E^{*}_{**}$ collapses for $F = Q$. For then, using a Bockstein spectral sequence argument to obtain $\operatorname{rank} H^{n}(Y, Z_p) \geq \operatorname{rank} H^{n}(Y,Q)$ and one has:

$$\begin{aligned}
\operatorname{rank} H^{n}(Y,Q) &= \sum_{q+q'=n} \operatorname{rank} {}^{Q}E^{\infty}_{q,q'} = \sum_{q+q'=n} \operatorname{rank} {}^{Q}E^{2}_{q,q'} = \\
&= \sum_{q+q'=n} \operatorname{rank} \operatorname{Tor}^{H^*(X_2,Q)}_{q,q'}(H^*(X_1, Q), Q) = \\
&= \sum_{q+q'=n} \operatorname{rank} \operatorname{Tor}^{H^*(X_2,Z_p)}_{q,q'}(H^*(X_1, Z_p), Z_p) = \\
&= \sum_{q+q'=n} \operatorname{rank} {}^{Z_p}E^{2}_{q,q'} \geq \sum_{q+q'=n} \operatorname{rank} {}^{Z_p}E^{\infty}_{q,q'} = \\
&= \operatorname{rank} H^{n}(Y, Z_p) \geq \operatorname{rank} H^{n}(Y,Q)
\end{aligned}$$

Hence all the (weak) inequalities are actually equalities and ${}^{Z_p}E^{2} = {}^{Z_p}E^{\infty}$. Moreover, $\operatorname{rank} H^{n}(Y, Z_p) = \operatorname{rank} H^{n}(Y,Q)$ implies the collapse of the Z_p - Bockstein s.s and $H^*(Y,Z)$ is p-torsion free.

The collapse of the Eilenberg-Moore spectral sequence for $F = Q$ one can see by comparing

5.6.2.1.
$$\begin{array}{ccc}
Y & \xrightarrow[\approx 0]{} & \hat{K}(N_2) \times \Omega\hat{K}(M_2) \\
\downarrow & & \downarrow j \\
X_1 & \xrightarrow[\approx 0]{} & \hat{K}(N_1) \times \hat{K}(N_2) = K(X_1) \\
\downarrow & & \downarrow \hat{K}(A(f)) \\
X_2 & \xrightarrow[\approx 0]{} & \hat{K}(M_1) \times \hat{K}(M_2) = K(X_2)
\end{array}$$

where $\hat{K}(N_i) = K(QN_i)$, $H^*(\hat{K}(N_i, Z), Z)/\text{torsion} = N_i$ and similarly for $\hat{K}(M_i)$.

To see that fiber $\hat{K}(A(f)) = K(N_2) \times \Omega\hat{K}(M_2)$ observe

$$\begin{array}{ccccc}
\hat{K}(N_2) \times \Omega\hat{K}(M_2) & \longrightarrow & \hat{K}(N_2) & \xrightarrow{\quad * \quad} & \hat{K}(M_2) \\
\| & & \downarrow & & \downarrow \\
\text{fiber } K(A(f)) & \longrightarrow & \hat{K}(N_1) \times \hat{K}(N_2) & \xrightarrow{\hat{K}(A(f))} & \hat{K}(M_1) \times \hat{K}(M_2) \\
 & & \downarrow p_1 & \swarrow p_1 & \\
 & & \hat{K}(N_1) & = & \hat{K}(M_1)
\end{array}$$

The collapse of the s.s for the right hand side fibration of 5.6.2.1 is obvious.

If one applies 5.6.2 for $X_1 = *$ one gets:

5.6.3. Corollary:

(a) *Let* X *be a space with* $H_*(X,Z)$ p-*torsion free and* $H^*(X, Z_p)$ *a free algebra. Then* $H_*(\Omega X, Z)$ *is* p-*torsion free and* $H^*(\Omega X, F) = \mathrm{Tor}^{H^*(X,F)}(F,F)$, $(F = Z_p$ *or* $Q)$ *as coalgebras.*

(b) *If* X_1, X_2, f, Y *are as in* 5.6.2 *then*

$$H^*(Y,F) \approx H^*(X_1,F)//\mathrm{im}\, H^*(f,Z) \otimes \mathrm{im}\, H^*(j,F)$$

as an $H^*(X_1, F)//\mathrm{im}\, H^*(f,Z)$ *module and* $\mathrm{im}\, H^*(j,F)$ *comodule,* $j\colon \Omega X_2 \to Y$ *from the Puppe sequence*

$$\Omega X_1 \xrightarrow{\Omega f} \Omega X_2 \xrightarrow{j} Y \longrightarrow X_1 \xrightarrow{f} X_2.$$

If in addition $H^*(\Omega X_2, F)$ *is a free algebra (always holds for* $F = Q$) *then the above representation of* $H^*(Y,F)$ *is an algebra representation.*

5.6.4. Let X be a space, $H^*(X,Z)$ p-torsion free and $H^*(X, Z_p)$ a free algebra. We construct a BP Adams resolution for X as follows:

Let $f_0\colon X \to E_0(X) = \prod_i B(n_i, p)$ yield a surjection $H^*(f_0, Z_p)$ (and obviously of $H^*(f_0, Q)$). Let $A = \ker QH^*(f_0, Z_p)$. Lift $\hat{k}_1\colon E_0(X) \to K(A)$ to $k_1\colon E_0(X) \to FK(A) = \prod_j B(m_j, p)$, k_1 can be chosen so that $k_1 f_0 \sim *$.

Indeed, $\hat{k}_1$ factors through $C(f_0)$ - the cone on f_0. As $H^*(f_0,Z)$ is a p-surjection $H^*(C(f_0), Z) \to H^*(E_0(X), Z)$ is p-injective and $H^*(C(f_0), Z)$ is p-torsion free. One can lift $C(f_0) \to K(A)$ to $C(f_0) \to F(K(A))$ which induces k_1, k_1 factors through $C(f_0)$.

Suppose a resolution $E_i(X)$, f_i $i \leq n$ exists, $H^*(E_n(X), Z)$ p-torsion free, $H^*(E_n(X), Z_p) \approx H^*(X, Z_p) \otimes M_n$, M_n - a free algebra, M_n generating the ideal $\ker H^*(f_n, Z_p)$. Map $\hat{k}_n\colon E_n(X) \to K(QM_n)$, $\hat{k}_n f_n \sim *$. Lift $\hat{k}_n$ to $k_n\colon E_n(X) \to F(K(QM_n))$ so that $k_n f_n \sim *$ (which can be done as $H^*(f_n, Z)$ is p-surjective). Let $E_{n+1}(X) = \text{fiber } k_n$. Lift f_n to $f_{n+1}\colon X \to E_{n+1}(X)$. The cohomology of $E_{n+1}(X)$ can be obtained by 5.6.2, 5.6.3. (as $H^*(F(K(QM_n)) = \prod_i B(n_i, p), Z_p)$ is a free algebra). $H_*(E_{n+1}(X), Z)$ is p-torsion free and

$$H^*(E_{n+1}(X), Z_p) \approx H^*(X, Z_p) \otimes M_{n+1},$$

$$M_{n+1} = \operatorname{im} H^*(E_{n+1}(X), Z_p) \to H^*(\Omega F(K(QM_n)), Z_p)$$

Finally we shall outline the construction of a BP Adams resolution for an H-space X with $H_*(X,Z)$ p-torsion free, $H^*(X,Q)$ primitively generated and $H_*(X, Z_p)$ a free algebra:

One starts with an H-map

$$f_0 \colon X \to \prod_{j \in J_0} B(n_j, p) = E_0(X), \quad H^*(f_0, Z) \text{ p-surjection}$$

This is done by lifting $X \to K(PH^*(X,Z)/\text{torsion}) = K_0$ to an H-map $f_0 \colon X \to \Omega F(BK_0) = \prod_{j \in J_0} B(n_j, p)$ (BK_0 - the classifying space of K_0, $\Omega BK_0 = K_0$). As $H_*(\Omega F(BK_0), Z)$ is a free algebra one can see that $H^*(f_0, Z)$ is p-surjective. Let $A \subset H^*(E_0(X), Z_p)$ be a sub-Hopf algebra generating $\ker H^*(f_0, Z_p)$. Realize A by an H-map $k_1 \colon E_0(X) \to \hat{B}_1(X) = \prod_{j \in J_1} B(n_j, p)$, $k_1 f_0 \sim *$. Let $E_1(X) = \text{fiber } k_1$ lift f_0 to an H-map $f_1 \colon X \to E_1(X)$. Indeed, if $B_1(X) = \Omega\hat{B}_1(X)$ then

$$HD(f_1) \in [X \wedge X, E_1(X)]$$

is in $\operatorname{im}[X \wedge X, B_1(X)] \to [X \wedge X, E_1(X)]$ and as $X \wedge X \to E_1(X) \wedge E_1(X)$ yields a mod p surjection of integral cohomology by 5.3.5 (c) $[E_1(X) \wedge E_1(X), B_1(X)] \to [X \wedge X, B_1(X)]$ is surjective and one can alter the multiplication in $E_1(X)$ by a map $E_1(X) \wedge E_1(X) \to B_1(X) \to E_1(X)$ so that f_1 is an H-map. $H^*(E_1(X), Z_p) \approx H^*(X, Z_p) \otimes A_1$, $A_1 \approx \operatorname{im}[H^*(E_1(X), Z_p) \to H^*(B_1(X), Z_p)]$ as Hopf algebras, as $X \to E_1(X)$ is an H-map and $j_1 \colon B_1(X) \to E_1(X)$ is central. $A_1 \approx \ker H^*(f_1, Z_p)$ is a free algebra and one can proceed by realizing A_1 by $E_1(X) \to \hat{B}_2(X) \approx \prod_{j \in J_2} B(n_j, p)$ etc.

5.7. Some simple applications.

The simplicity referred to in the title of this section has to do only with the fact that we use mostly 3-stage BP resolutions in our computations. The arithmetical calculations however may be complex.

5.7.1. Proposition: *Let* X *be an* H_0 *space,* $H_*(X,Z)$ *p-torsion free. If* ${}^p\pi_{n_0}(X) \neq 0$, ${}^p\pi_n(X) = 0$ *for* $n < n_0$ *then either* $QH^{n_0+1}(X, Z_p) \neq 0$ *or* $QH^{n_0-2p+3}(X, Z_p) \neq 0 \neq QH^{n_0-2p+3}(X,Q)$.

Proof: Map $f_0: X \to B_0(X) = \prod_{i \in J} B(n_i, p)$ so that $H^*(f_0, Z_p)$ is p-surjective. We may assume that f_0 is efficient in the sense that the image of the composition $H^{n_i}(B(n_i, p), Z_p) \to H^{n_i}(B_0(X), Z_p) \to H^{n_i}(X, Z_p)$ is not in the image of $H^{n_i}(\prod_{j<i} B(n_j,p),Z_p) \to H^{n_i}(B_0(X),Z_p) \to H^{n_i}(X,Z)$. Note that $QH^m(B(n_i, p), Z_p) = 0$ unless $m \equiv n_i \ (2p-2)$ and that $QH^{n_i+2k(p-1)}(B(n_i, p), Z_p) \neq 0$ for all $k \geq 0$. Let m_0 be the minimal integer with $\ker H^{m_0}(f_0, Z_p) \neq 0$. Let $A = \ker H^{m_0}(f_0, Z)/\text{torsion}$ and let $\hat{k}_1: B_0(X) \to K(A, m_0)$ realize A. As X is an H_0 space $QH^{m_0}(\hat{k}_1, Q) \neq 0$. $\hat{k}_1 f_0 \sim *$. If $E_1(X) = \text{fiber } \hat{k}_1$ then f_0 lifts to $f_1: X \to E_1(X)$ and $H^*(f_1, Z_p)$ is an isomorphism in $\dim \leq m_0$, hence ${}^p\pi_n(f_1)$ is an isomorphism for $n < m_0$. Obviously ${}^p\pi_n(E_1(X)) = 0$ for $n < m_0-1$. We shall show now that ${}^p\pi_{m_0-1}(E_1(X)) \neq 0$, as a matter of fact ${}^p\pi_{m_0-1}(E_1(X)) \otimes Z_p = A \otimes Z_p$:

$\hat{k}_1$ obviously factors as $B_0(X) \to \prod_{n_i \le m_0} B(n_i,p) \xrightarrow{k_1} K(A, m_0)$

Now $\prod_{n_i = m_0} B(n_i, p) \xrightarrow{i_1} \prod_{n_i \le m_0} B(n_i, p) \xrightarrow{k_1} K(A, m_0)$ must yield the zero map on Z_p cohomology by the efficiency of f_0. Hence $k_1 i_1$ is divisible by p and $\pi_{m_0}(k_1 i_1) \otimes Z_p = 0$. Now, for any map $k_1^{(i)}: B(n_i, p) \to K(A, m_0)$, if $n_i < m_0$ then $\pi_{m_0}(k_1^{(i)}) \otimes Z_p = 0$ as the Hurewicz homomorphism $\sigma_n: \pi_n(B(n_i, p)) \to H_n(B(n_i, p), Z)$ $n > n_i$ satisfies $\sigma_n \otimes Z_p = 0$ (as 0-mod p k-invariants will violate the minimality of $B(n_i, p)$ - 5.2.2 (e)). Hence,

$$\pi_{m_0}(k_1) \otimes Z_p = \pi_{m_0}(k_1 i_1) \otimes Z_p \oplus \Big(\sum_{n_i < m_0} \pi_{m_0}(k_1^{(i)}) \otimes Z_p \Big) = 0.$$

Now one can easily see that $\pi_{m_0 - 1} E_1(X) = \operatorname{coker} \pi_{m_0}(k_1)$, $\pi_{m_0-1}(E_1(X)) \otimes Z_p = \operatorname{coker} \pi_{m_0}(k_1) \otimes Z_p = A \otimes Z_p$. Now, if for no $j \in J$ $n_j = m_0$, $(k_1 i_1 = *)$, $QH^{m_0}(k_1, Q) \ne 0$ implies $QH^{m_0}(\prod_{i \in J} B(n_i, p), Q) \ne 0$, consequently for some $i \in J$ $m_0 \equiv n_i$ $(2p-2)$, $n_i < m_0$, but then $QH^{m_0 - 2p+2}(X, Z_p) \ne 0 \ne QH^{m_0 - 2p+2}(X, Q)$. If for some $j \in J$ $n_j = m_0$ as $QH^{m_0}(k_1 i_1, Z_p) = 0$ implies $QH^{m_0}(B(m_0, p), Z_p) \xrightarrow{\ne 0} QH^{m_0}(X, Z_p)$. Finally observe that $n_0 + 1 = m_0$.

<u>5.7.2. Proposition</u>: *Let* X, μ *be an* H-space, $H^*(X,Q)$ *primitively generated,* $H^*(X, Z_p)$ - *exterior algebra on odd dimensional generators. Suppose* X *is* $2n$ *connected* mod p *(i.e.* $H^{2n+1}(X, Z_p) \ne 0$ *and*

$H^m(X, Z_p) = 0 \quad m \leq 2n$). *Then*

$$Ht_{a(n,p)}(X) \approx \prod_{i=1}^{p-1} X_i, \text{ where } X_i \text{ satisfy:}$$

for $m \leq a(n,p) + 1$ $PH^m(X_i, Z_p) \neq 0$ *only if* $m \equiv 2i+1 \ (2p-2)$. $a(n,p)$ *is given by*

$$a(n,p) = \begin{cases} 2np + 2p-4 & n \leq p-1 \\ 2n + 2p(p-1)-2 & n \geq p-1 \end{cases}$$

<u>Proof</u>: In this range X is a two stage BP resolution. One can resolve $PH^*(X, Z_p)$ in $\dim \leq a(n,p) + 2$ as follows:

$$PH^*(X, Z_p) \text{ is a } Z_p[P^1] \text{ module}, \ P^1 \in A(p)$$

Hence, $PH^*(X, Z_p) = \bigoplus_j A_j$, $A_j = [Z_p[P^1]/(P^1)^{r_j}](a_j)$ $r_j < p$. Realize every $a_j \in PH^{2n_j+1}(X, Z_p)$ by $f_{0,j}: X \to B(2n_j+1, p)$, $f_{0,j}$ - H-map. Put $f_0: X \to B_0(X) = \prod_j B(2nj+1, p)$. Note that

$$PH^*(B(2n_j+1, p), Z_p) = Z_p[P^1](v_j)$$

in $\dim \leq a(n,p) + 2$. Let

$$\hat{k}_{1,j}: B(2n_j+1, p) \to K(Z_p, 2n_j + 2r_j(p-1)),$$

$[k_{1,j}] = (P^1)^{r_j} = \frac{1}{r_j} P^{r_j}$. Then $k_{1,j} f_0 \sim *$. Put $B_{0,i}(X) = \prod_j B(2n_j+1, p)$, $K_i = \prod_j K(Z_p, 2n_j + 2r_j(p-1) + 1)$ products taken over all j with $n_j \equiv i$ mod$(p-1)$. Lift $\prod_j k_{1,j} = \tilde{k}_{1,i}: B_{0,i}(X) \to K_i$ to an H-map

$k_{1,i}: B_{0,i}(X) \to F(K_i) = B_{1,i}(X)$, then as

$$PH^*(f_0, Z_p)[PH^{2i+1+2k(p-1)}(B_{0,i}(X), Z_p)] = PH^{2i+1+2k(p-1)}(X, Z_p)$$

one can choose $k_{1,i}$ so that $k_{1,i}f_0 \sim *$. Put $E_i(X) = \text{fiber } k_{1,i}$ then f_0 lifts to $f_1: X \to \prod_i E_i(X)$ which is a p-equivalence in $\dim \leq a(n,p)$. Put $X_i = Ht_{a(n,p)}(E_i(X))$ and then $Ht_{a(n,p)}(X) \approx_p \prod X_i$ is the desired decomposition.

To illustrate the use of the BP resolutions for specific computations we have:

<u>5.7.3. Proposition</u>: *If* $4 \leq m < 2p^2 - 2$ *then*

$$_p\pi_m(S^3) = \begin{cases} 0 & \textit{if} \quad m \not\equiv 1, \ 2 \mod 2p-2 \\ 0 & \textit{if} \quad m = 2p-1 \\ Z_p & \textit{otherwise} \end{cases}$$

<u>Proof</u>: Consider the resolution:

$$B(3,p) \xrightarrow{k_0} B(2p+1, p) \times B(4p-1, p) \xrightarrow{k_1} B(4p-1, p).$$

k_0 and k_1 have the properties: $H^*(k_0, Z_p)(u_{2p+1} \otimes 1) = P^1 u_3$. $0 \neq H^*(k_0, Z_p)(1 \otimes u_{4p-1}) \in PH^{4p-1}(B(3,p), Z_p)$, $H^*(k_1, Z_p)u_{4p-1} = P^1 u_{2p+1} \otimes 1$ ($u_i \in PH^i(B(i,p), Z_p)$ - a generator). k_0, k_1 are H-maps and $k_1 k_0 \sim *$. k_i give rise to the following tower:

$$\begin{array}{ccccc}
S^3 & \xrightarrow{h_2} & E_2 = \text{fiber } \hat{k}_1 & & \\
& & \downarrow & & \\
\Omega B_0(S^3) & \xrightarrow{\hat{j}_1} & E_1 = \text{fiber } k_0 & \xrightarrow{\hat{k}_1} & B(4p-2,\ p) = \Omega B(4p-1,\ p) \\
& & \downarrow & & \\
& & E_0 = B(3,p) & \xrightarrow{k_0} & B_0(S^3) = B(2p+1,p)\times B(4p-1,p)
\end{array}$$

$\hat{k}_1\hat{j}_1 = \Omega k_1$.

$H^s(h_2, Z_p)$ is an isomorphism for $s \leq 2p^2-2$ hence ${}^p\pi_m(h_2)$ is an isomorphism for $m < 2p^2-2$. It can be seen that

$$ {}^p\pi_m(E_2) = \begin{cases} 0 & m \not\equiv 1,\ 2,\ (2p-2) \\ {}^p[\operatorname{coker} \pi_{m+1}(k_0)] & m \equiv 2 \quad (2p-2) \\ {}^p[\operatorname{coker} \pi_{m+2}(k_1)] & m \equiv 1 \quad (2p-2) \end{cases} $$

To compute $\operatorname{coker}(\pi_*(k_\varepsilon))$ consider $k_{0,2}\colon B(3,p) \to B(4p-1)$. By 5.5.3, $B(3,p)$ is a mod p retract of SU, $PH^*(k_{0,2}, Z_p)u_{4p-1} = w_{4p-1} \neq 0$ and as $P^j w_{4p-1} = \binom{2p-1}{j} w_{4p-1+2j(p-1)}$, $\binom{2p-1}{j} \neq 0 \quad j \leq p$,

$QH^*(k_{0,2}, Z_p) = PH^*(k_{0,2}, Z_p)$ is an isomorphism in dim b, $4p-1 \leq b \leq 2p^2 + 2p-1$. Hence for $2p-1 \leq s \leq p^2 + p-1$, $s = 1 + k(p-1)$, $2 \leq k \leq p$, one has

$$\begin{array}{ccc}
\pi_{2s+1}(B(3,p)) & \xrightarrow{\pi_{2s+1}(k_{0,2})} & \pi_{2s+1}(B(4p-1,\ p)) \\
\Big\downarrow \sigma_{2s+1} & & \Big\downarrow \hat{\sigma}_{2s+1} \\
PH_{2s+1}(B(3,p),Z) & \xrightarrow[H_{2s+1}(k_{0,2},Z)]{\approx\ p} & PH_{2s+1}(B(4p-1,\ p)
\end{array}$$

and degree $\pi_{2s+1}(k_{0,2}) = [\text{degree } \sigma_{2s+1}/\text{degree } \hat{\sigma}_{2s+1}] \cdot c, \quad c \in Z,$ $(c,p) = 1$.

By a theorem of Bott (and as $B(3,p)$ is a mod p retract of SU) degree $\sigma_{2s+1} = s!$ and by the Brown Peterson construction degree$(\hat{\sigma}_{2s+1}) = p^{k-2} \cdot c_1$, $(c_1,p) = 1$ and k is given by $s-1 = k(p-1)$. Degree $\pi_{2s+1}(k_{0,2})$ is then

$$c_2 \cdot [1 + k(p-1)]!/p^{k-2} = c_2[(k-1)p + p - k+1]!/p^{k-2}.$$

Now $((k-1)p + p - k+1)! = c_3(k-1)!p^{k-1} \quad 1 < k \leq p-1$. Hence, degree $\pi_{2s+1}(k_{0,2}) = c\cdot p$, ${}^{p}[\text{coker } \pi_{2s+1}(k_{0,2})] = Z_p$.

To compute degree $\pi_{2s+1}(k_{0,1})$, $k_{0,1}\colon B(3,p) \quad B(2p-1,\ p)$ consider

$$\Omega^{2p-2}(k_{0,2})\colon \Omega^{2p-2}(B(3,p)) \to \Omega^{2p-2}B(4p-1,\ p)$$

As $\Omega B(3,p) = K(Z,2) \times B(2p,p)$, $\Omega^{2p-2}B(3,p) = \Omega^{2p-3}B(2p,p) = B(3,p)$, $\Omega^{2p-2}B(4p-1) = B(2p+1)$. Now $(\sigma)^{2p-2}\colon PH^{m}B(3,p) \to PH^{m-2p+2}B(3,p)$ is an isomorphism for $m = 4p-1$ (as no k $2p+1 \leq k \leq 4p-1$ is divisible by $2p$). Hence, $H^*(k_{0,1},\ Z_p) = H^*(\Omega^{2p-2}k_{0,2})$, degree $\pi_{2s+1}(k_{0,1})$ = degree $\pi_{2s+2p-1}(k_{0,2}) = p\cdot c$ for $s = 1 + k(p-1)$ $1 \leq k \leq p-1$. To compute $\pi_{2s+1}(k_1)$ put $k_{1,1}\colon B(2p+1, p) \to B(4p-1, p)$, $k_{1,1} = k_1 \cdot i_1$, $H^*(k_{1,1}, Z_p)u_{4p-1} = u_{2p+1}$, hence, $QH^i(k_{1,1}, Z_p)$ is an isomorphism for $4p-1 \leq i \leq 2p^2$. Thus, for $2 \leq k \leq p-1$

$$\begin{array}{ccc}
\pi_{3+2k(p-1)}(B(2p+1,p)) & \xrightarrow[\pi_*(k_{1,1})]{} & \pi_{3+2k(p-1)}(4p-1,p) \\
\Big\downarrow \sigma_k & & \Big\downarrow \hat{\sigma}_k \\
PH_{3+2k(p-1)}(B(2p+1,p),Z) & \xrightarrow[\approx\, p]{} & PH_{3+2k(p-1)}((4p-1,p),Z)
\end{array}$$

Degree $\sigma_k = c_1 \cdot p^{k-1}$, degree $\hat{\sigma}_k = c_2 p^{k-2}$, hence,

degree $\pi_{3+2k(p-1)}(k_{1,1}) = \hat{c}\cdot p$, $2 \leq k \leq p-1$. As $k_1 k_0 \sim *$

$$\text{degree } \pi_*(k_{0,1})\cdot\text{degree } \pi_*(k_{1,1}) + \text{degree } \pi_*(k_{0,2})\cdot\text{degree } \pi_*(k_{1,2}) = 0$$

Consequently; degree $\pi_{2s+1}(k_{1,2}) = c'\cdot p$ $(k_{1,2}\colon B(4p-1, p) \to B(4p-1, p))$.

Hence, ${}^p[\text{coker } \pi_{2s+1}(k_0)] = Z_p$ for $s \geq p$, $s \equiv 1$, mod $(p-1)$ and ${}^p[\text{coker } \pi_{2s+1}(k_1)] = Z_p$ for $s \geq 2p-1$ $s \equiv 1$ $(p-1)$. 5.7.3 follows.

A more technical computation but along the same lines leads to

<u>5.7.4. Proposition</u>: *Given integers* $n, m, n \geq p-1, m \leq p(p-1) - 1$. *Define* $j, k, a, \bar{a}$ *by* $m = j + k(p-1)$, $0 \leq j < p-1$, $n+j = ap + \bar{a}$, $0 \leq \bar{a} < p$. *Then*

$$^{p}\pi_{2n+2m}(SU(n)) = Z_{p^{r(n,m)}}$$

$$^{p}\pi_{2n+2m-1}(SU(n)) = Z_{p^{s(n,m)}}$$

where

$$p^{r(n,m)} = \begin{cases} \gamma_p[(a+k)!]p^{a} & \text{if } k \le \bar{a} \\ \min(\gamma_p[(a+k)!]p^{a},\ \gamma_p[(a+k-1)!]\cdot p^{a+\bar{a}}), & \text{if } k > \bar{a} \end{cases}$$

$$p^{s(n,m)} = \begin{cases} 1 \quad (\text{i.e. } Z_{p^{s(n,m)}} = 0) & \text{if } k \le \bar{a} \\ \min(p^{\bar{a}+1},\ \gamma_p[(a+k)]\cdot p & \text{if } k > \bar{a} \end{cases}$$

where $\gamma_p(b)$ *is the highest power of* p *dividing* b. *(Compare this with* $[\textit{Krishnarao}]_{1,2}$*).*

Outline of computations: The Adams resolution in this range has just 3 stages and is given by: $SU(n) \xrightarrow{f_0} SU \xrightarrow{k_0} B_0 \xrightarrow{k_1} B_1$, $B_0 = B_0' \times B_1$, $B_0' = \prod_{\hat{j}=0}^{p-2} B(\alpha_{\hat{j}}, p)$, $B_1 = \prod_{\hat{j}=0}^{p-2} B(\gamma_{\hat{j}}, p)$, $\alpha_{\hat{j}} = 2n+2\hat{j}+1$, $\gamma_{\hat{j}} = 2n+2\hat{j}+1+2(p-1)(\bar{a}(\hat{j})+1)$ where $\bar{a}(\hat{j})$ is defined for every integer $\hat{j}$ by $n+\hat{j} = a(\hat{j})p + \bar{a}(\hat{j})$, $0 \le \bar{a}(\hat{j}) < p$. $k_{0,1} = p_1k_0\colon SU \to B_0'$ is given by $PH^{\alpha_{\hat{j}}}(k_{0,1}, Z_p)$ is an isomorphism, $0 \le \hat{j} \le p-2$, $k_{0,2} = p_2k_0\colon SU \to B_1$ is given by $PH^{\gamma_{\hat{j}}}(k_{0,2}, Z_p)$ is an isomorphism $0 \le \hat{j} \le p-2$, k_1 is given by $k_1i_1\colon B_0 \to B_1$ satisfies $PH^{\gamma_{\hat{j}}}(k_{0,2}, Z_p)u_{\gamma_{\hat{j}}} = p^{\bar{a}(j)+1}u_{\alpha_{\hat{j}}}$, k_i H-maps and $k_1k_0 \sim *$. As in 5.7.3 here too we have

$$^p\pi_{2n+2s}(SU(n)) = {}^p[\text{coker}\ \pi_{2n+2s+1}(k_0)]$$

$$^p\pi_{2n+2s-1}(SU(n)) = {}^p[\text{coker}\ \pi_{2n+2s+1}(k_1)].$$

and to compute these one has to compute degree $\pi_*(k_{i,j})$ $i = 0,1$, $j = 1,2$. degree $\pi_*(k_{0,1})$ is computed by delooping $SU \to B(\alpha_j, p)$ $2(p-\bar{a}-1)$ times to obtain maps $SU \to B(2n+2j+1 + 2(p-\bar{a}-1), p)$ with $PH^*(\ \ , Z_p)$ an isomorphism in the range of interest. One obtains

$$^p[\text{coker}\ \pi_{2n+2m+1}(k_{0,1})] = Z_{p^{r_1(n,m)}},$$

$$p^{r_1(n,m)} = \gamma_p[(a+k)!]p^a.$$

A direct computation yields

$$^p[\text{coker}\ \pi_{2n+2m+1}(k_{0,2})] = Z_{p^{r_2(n,m)}}$$

$$r_2(n,m) = 0 \qquad \text{if}\ \ k \le \bar{a},$$

$$p^{r_2(n,m)} = \gamma_p[(a+k-1)!]p^{a+\bar{a}}, \qquad \text{if}\ \ k > \bar{a}$$

$$^p[\text{coker}\ \pi_{2n+2m+1}(k_{1,1})] = Z_p\bar{a}+1$$

$k_1k_0 \sim *$ yields for

$$d = \text{degree}\ \pi_{2n+2m+1}(k_{1,2}):\ p^{r_1(n,m)} \cdot p^{\bar{a}+1} = p^{r_2(n,m)}\gamma_p(d),$$

hence, $\gamma_p d = \gamma_p(a+k)\cdot p$. 5.7.4 now follows.

As a final application of the BP resolutions we have:

5.7.5. Proposition: *Given a* $4n+2$ *dimensional space* Y. *Suppose* $H^{4n-1}(Y,Z) = 0 = H^{4n+1}(Y,Z)$ *and* $H^*(Y,Z)$ *is* 2 *torsion free in* $\dim \geq 4n-2$ *and* 3 *torsion free in* $\dim \geq 4n$. *Then for any complex vector bundle* ξ *of (complex) dimension* $2n+1$ *the obstruction for* ξ *to have* 4-*sections are the* 4 *top integral Chern classes of* ξ.

Proof: The obstructions in question are the obstructions in the lifting problem:

$$\begin{array}{ccc} & & BU(2n-3) \\ & \nearrow & \big\downarrow h \\ Y & \xrightarrow{\ f\ } & BU(2n+1) \end{array}$$

(where f classifies ξ).

To study the lifting problem one prime at a time note that if X_i, $i = 1,\ldots,k$, are spaces so that for every i h factors as

$$BU(2n-3) \xrightarrow{h_i'} X_i \xrightarrow{h_i''} BU(2n+1), \quad h_i''h_i' \sim h.$$

h_i' - a mod $\mathbb{P}_i$ equivalence in $\dim \leq 4n+2$ and $\{\mathbb{P}_i\}$ $i = 1,\ldots,k$ is a partition of $\mathbb{P}$, then $f: Y \to BU(2n+1)$ lifts to $BU(2n-3)$ if and only if f lifts to $\tilde{f}_i: Y \to X_i$ for every i.

Let $\mathbb{P}_1 = \{2\}$, $\mathbb{P}_2 = \{3\}$, $\mathbb{P}_3 = \mathbb{P} - \{2,3\}$.

Let X_3 be the fiber of $w: BU(2n+1) \to \prod_{j=0}^{3} K(Z, 4n-4+2j)$ realizing the Chern classes. Then h factors $BU(2n-3) \xrightarrow[\leq 4n+2]{\approx \mathbb{P}_3} X_3 \to BU(2n+1)$

and the obstructions of an $\tilde{f}_3$ lifting are obviously the Chern classes.

To construct X_1 we distinguish between two cases.

If n is odd, consider

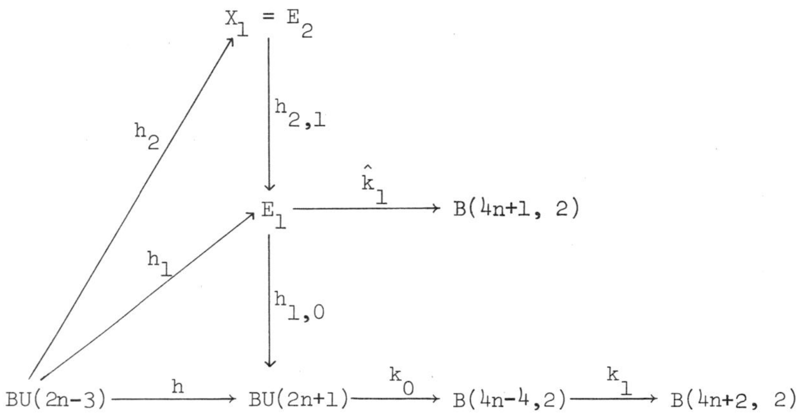

$H^{4n-4}(k_0, Z) = w_{4n-4}$, $k_0 h \sim *$. Such a k_0 exists as $H^*(h,Z)$ is surjective. $H^*(k_0, Z_2)$ is surjective in dim $< 4n+4$. k_1 is given by $H^*(k_1,Z_2)u_{4n+2} = Sq^{4,2}u_{4n-4}$. $B(4n-5,2) = \Omega B(4n-4, 2) \to E_1 \xrightarrow{\hat{k}_1} B(4n+1, 2)$ is homotopic to Ωk_1. By 5.3.1 $\hat{k}_1 h_1 \sim *$. $QH^*(\hat{k}_1, Z_2)$ is a monomorphism through dim $4n+4$, hence h_2 is a mod 2 equivalence through dim $4n+2$.

For n even the resolution is given by

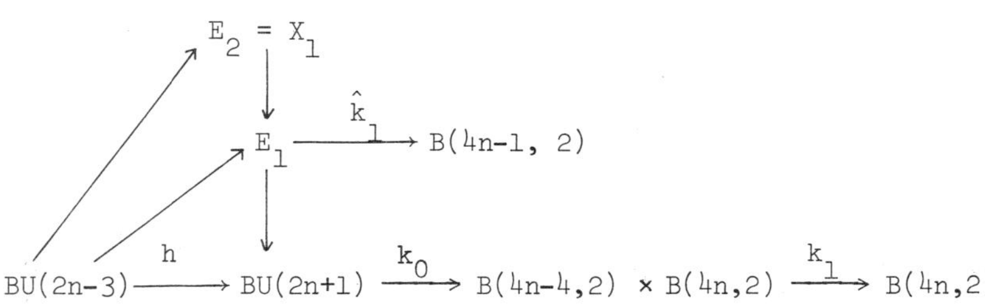

$H^*(k_0, Z)u_{4n-4} \otimes 1 = w_{4n-4}, \quad H^*(k_0, Z)1 \otimes u_{4n} = w_{4n}$

$H^*(k_1, Z_2)u_{4n} = Sq^4 u_{4n-4} \otimes 1.$

As $H^*(Y,Z)$ is 2-torsion free in $\dim > 4n-4$ $k_0 f \sim *$ if and only if $H^*(k_0 f, Q) = 0$. As $\mathrm{im}H^{4n-4+2j}(k_0, Q)$ are the Cherm classes f lifts to E_1 and as by hypothesis $H^{4n-1}(Y,Z) = 0 = H^{4n+1}(Y,Z)$ f lifts further to $\tilde{f}_1 : Y \to X_1$.

Similar arguments hold for X_2. Here we have three cases:

$2n \equiv 1 \mod 3$

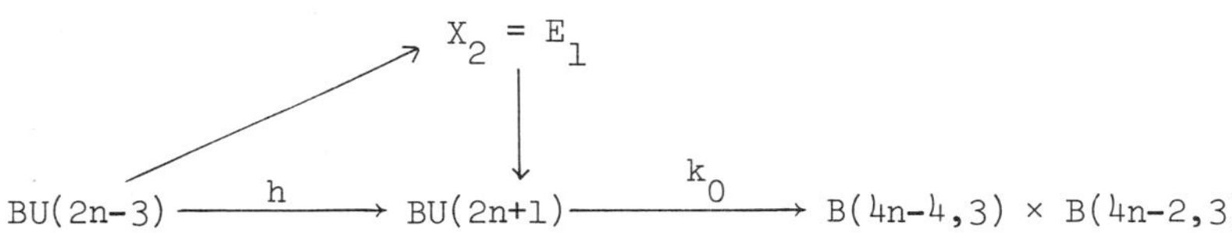

$w_{4n-2a} \in \mathrm{im}H^{4n-2a}(k_0, Z) \qquad a = 1, 2.$

$2n \equiv 2 \mod 3$

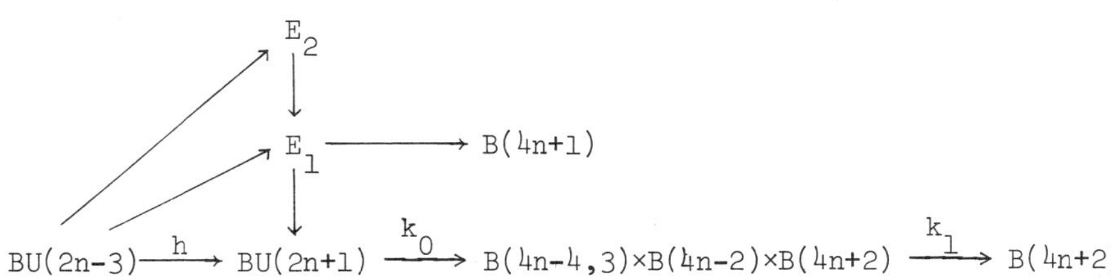

$H^{4n+1}(k_1, Z_3)\hat{u}_{4n+2} = 1 \otimes \mathcal{P}^1 u_{4n-2} \otimes 1.$

$2n \equiv 0 \mod 3$

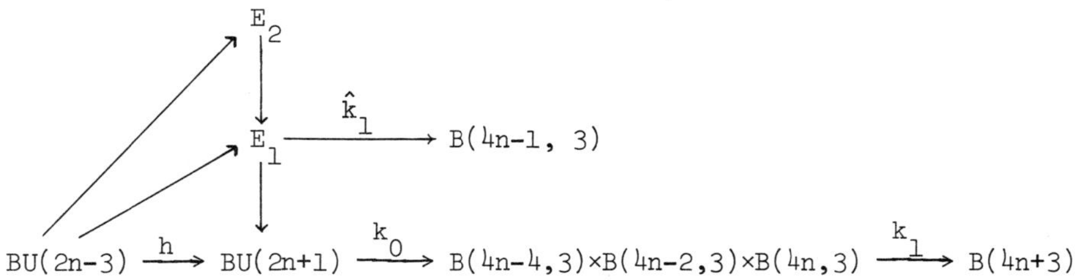

Bibliography

Adams, J.F.,

1. On the non existence of elements of Hopf invariant one. Ann. of Math. 72 (1960), 20-104.

2. The sphere considered as an H-space mod p. Quart. J. Math., Oxford (2) 12 (1961), 52-60.

3. H-spaces with few cells. Topology 1 (1962), 67-72.

Araki, S. and Kudo T., Topology of H_n spaces and H-squaring operations. Mem. Fac. Sci. Kyusyu Univ., 10 (1956), 85-120.

Arkowitz, M., Homotopy products for H-spaces. Mich. Math. J. 10 (1963), 1-9.

Arkowitz, M. and Curjel, C.R.,

1. On the number of multiplications of an H-space. Topology 2 (1963), 205-209.

2. Homotopy commutators of finite order, II. Quart. J. Math., Oxford (2) 15 (1964), 316-326.

3. On maps of H-spaces. Topology 6 (1967), 137-148.

4. Some properties of the exotic multiplications on the three sphere. Quart. J. Math., Oxford (2) 20 (1969), 171-176.

Berstein, I. and Ganea, T., Homotopical nilpotency. Ill. J. of Math. 5 (1961), 99-130.

Bott, R.,

1. The space of loops on a Lie group. Michigan Math. J. 5 (1958), 35-61.

2. The stable homotopy of the classical groups. Ann. of Math. 70 (1959), 313-337.

Browder, W.,

1. Torsion in H-spaces. Ann. of Math. 74 (1961), 24-51.

2. Homotopy commutative H-spaces. Ann. of Math. 75 (1962), 283-311.

3. Higher torsion in H-spaces. Trans. Amer. Math. Soc. 108 (1963), 353-375.

4. On differential Hopf algebras. Trans. Amer. Math. Soc. 107 (1963), 153-176.

5. Homology rings of groups. Amer. J. of Math. 90 (1968), 318-333.

Brown, E.H. and Peterson, F.P., A spectrum whose Z_p cohomology is the algebra of reduced pth powers. Topology 5 (1966), 149-154.

Copeland, A.H. Jr.,

1. Binary operations on set of mapping classes. Mich. Math. J. 6 (1959), 7-23.

2. On H-spaces with two nontrivial homotopy groups. Proc. Amer. Math. Soc. 8 (1957), 184-191.

Croon, F.H., Homotopy type of total loop spaces. Portugal Math. 30 (1971), 11-16.

Curjel, C.R., On the H-space structures of finite complexes. Comment. Math. Helv. 43 (1968), 1-17.

Curtis, M.,

1. Finite dimensional H-spaces. Bull. Amer. Math. Soc. 77 (1971), 1-12 and 1120.

2. H-spaces mod p. II. H-spaces, Neuchatel Aôut 1970. Lecture notes in Math. 196, Springer, 11-19.

Curtis, M. and Mislin G.,

1. Two new H-spaces. Bull. Amer. Math. Soc. 76 (1970), 851-852.

2. H-spaces which are bundles over S^7. J. Pure and Appl. Algebra 1 (1971), 27-40.

Freyd, P.,

1. Stable homotopy. Proceedings of the conference on categorical algebra (La Jolla 1965), Springer (1966).

2. Stable homotopy II. AMS Proceedings of symposia in pure mathematics, Vol. XVII (1970), 161-183.

Fuchs, M., The functor [,Y] and loop fibrations I. Mich. J. of Math. 14 (1967), 283-287.

Gilbert, W.J., Homotopical nilpotence of the seven sphere. Proc. Amer. Math. Soc. 32 (1972), 621-622.

Harper, J.R.,

1. Homotopy groups of finite H-spaces. Comment. Math. Helv. 47 (1972), 311-331.

2. The mod 3 homotopy type of F_4. Localization in group theory and homotopy theory (Seattle 1974), Lecture notes in Math. 418, Springer (1974). 58-67.

3. H-spaces with torsion. (Mimeographed).

Harrison, J., Scheerer H., Zur homotopietheorie von abbildungen in homotopieassoziative H-raum. Archiv der Mathematik Vol. 23 (1972), 319-323.

Hilton, P., Homotopy theory and duality, Gordon and Breach Science Publishers, 1965.

Hilton P., Mislin G. and Roitberg, J., Localization of nilpotent groups and spaces. North Holland Mathematics Studies 15 (1975).

Hilton P. and Roitberg, J., On principal S^3 bundles over spheres. Ann. of Math. 90 (1969), 91-107.

Hu. S., Homotopy theory, Academic Press, NY, 1959.

Hubbuck, J.R.,

1. Some results in the theory of H-spaces. Bull. Amer. Math. Soc. 74 (1968), 965-967.

2. On homotopy commutative H-spaces. Topology 8 (1969), 119-126.

3. Generalized cohomology operations and H-spaces of low rank. Trans. Amer. Math. Soc. 141 (1969), 335-360.

4. Finitely generated cohomology Hopf algebras. Topology 9 (1970), 205-210.

James, I.M.,

1. Reduced product spaces. Ann. of Math. (2) 62 (1955), 170-197.

2. Multiplications on spheres II. Trans. Amer. Math. Soc. 84 (1957), 545-558.

3. On H-spaces and their homotopy groups, Quart. J. Math., Oxford Ser (2) 11 (1960), 161-179.

Krishnarao, G.V.,

1. Unstable homotopy of O(n). Trans. Amer. Math. Soc. 127 (1967), 90-97.

2. Nonstable homotopy groups of U(n). General topology relations to modern Analysis and Algebra. Proc. Kampur Topology conference 1968, 173-174 (1971).

Kumpel, P.G., Jr.

1. Lie groups and product of spheres. Proc. Amer. Math. Soc. (1965), 1350-1356.

2. On p-equivalences of mod p H-spaces. Quart. J. Math., Oxford II, 23 (1972), 173-178.

Larmore, L.L. and Thomas, E., Mapping into loop spaces and central group extensions. Math. Z. 128 (1972), 277-296.

Lin, J.P.,

1. H-spaces with finitely generated cohomology algebras. Bull. Amer. Math. Soc. 80 (1974), 1233-1230.

2. Torsion in H-spaces I., Ann. of Math. (to appear).

Liulevicious, A., The factorization of cyclic reduced powers by secondary cohomology operations. Memoirs AMS 42 (1962).

Massey, W.S., Peterson, F.P., The mod 2 cohomology structure of certain fiber spaces. Memoirs of the Amer. Math. Soc. 74 (1967).

May, P. and Zabrodsky, A., H*(Spin (n)) as a Hopf algebra. (To appear).

Milgram, J.R., The bar construction and abelian H-spaces, Ill. J. Math. 11 (1967), 242-250.

Milnor, J. and Moore J.C., On the structure of Hopf algebras. Ann. of Math. 81 (1965), 211-264.

Mimura, M.,

1. Quelques groupes d'homotopie metastable des espaces symmetriques Sp(n) et U(2n)/Sp(n). C.R. Acad. Sci. Paris 262 (1966), 20-21.

2. The homotopy groups of Lie groups of low rank. J. Math., Kyoto Univ. 6 (1967), 131-176.

Mimura, M., Oneill, R.C. and Toda, H., On p-equivalence in the sense of Serre. Japan J. Math. 40 (1971), 1-10.

Mimura, M. and Toda H.

1. Homotopy groups of SU(3), SU(4), Sp(2). J. Math. Kyoto Univ. 3 (1963/64), 217-250.

2. On p-equivalences and p-universal spaces. Comment. Math. Helv. 46 (1971), 87-97.

Mislin, G.,

1. H-spaces mod p I. H-spaces, Neuchatel Aout 1970. Lecture notes in Math. 196, Springer (1970), 5-10.

2. The genus of an H-space. Symp. on Algebraic Topology. Battelle Seattle 1971, Lecture Notes in Math. 249, Springer (1971), 75-83.

Naylor, C.M., Multiplications on SO(3), Mich. Math. J. 13 (1966), 27-31.

Porter, G.J.,

1. Homotopy nilpotency of S^3. Proc. Amer. Math. Soc. 15 (1964), 681-682.

2. On H-fibrations. Quart. J. Math., Oxford (2) 22 (1971), 23-31.

Peterson, F., The mod p homotopy type of BSO and F/PL. Bol. Soc. Math. Mexicana (2) 14 (1969), 22-27.

Sigrist, F., Groupes d'homotopie des variétés de Stiefel complexes. Comment. Math. Helv. 43 (1968), 121-131.

Smith, L., Lectures on the Eilenberg Moore spectral sequence. Lecture notes in Mathematics 134, Springer (1970).

Spanier, E.H., Algebraic Topology, McGraw Hill.

Stasheff, J.,

1. On extension of H-spaces, Trans. Amer. Math. Soc. 105 (1962), 126-135.

2. Homotopy associativity of H-spaces I and II. Trans. Amer. Math. Soc. 108 (1963), 275-292 and 293-312.

3. Manifolds of the homotopy type of (non Lie) groups. Bull. Amer. Math. Soc. 75 (1969), 998-1000.

4. H-spaces and classifying spaces, foundations and recent developments. Proc. Symp. Pure Math. XXII (1970), 247-272.

5. H-spaces from the homotopy point of view. Lecture notes in Mathematics. 161, Springer (1970).

6. Sphere bundles over spheres as H-spaces mod $p > 2$. Symp. on Algebraic Topology, Battelle Seattle 1971, Lecture notes in Math. 249, Springer, 106-110.

Steenrod, N.E. and Epstein, D.B.A., Cohomology operations. Ann. of Math. Studies 50. Princeton University Press, 1962.

Thomas, E.,

1. On the mod 2 cohomology of certain H-spaces. Comment. Math. Helv. 37 (1962/63), 132-140.
2. Steenrod squares and H-spaces. Ann. of Math. 77 (1963), 306-317.
3. Steenrod squares and H-spaces II. Ann. of Math. 81 (1965), 473-495.
4. Postnikov invariants and higher order cohomology operations. Ann. of Math. (2) 85 (1967), 184-217.

Whitehead, G.W., On mappings into group like spaces. Comment. Math. Helv. 28 (1954), 320-328.

Wilkerson, C., Genus and cancellation for H-spaces, Localization in group theory and homotopy theory (Seattle 1974). Lecture notes in Math. 418, Springer. 157-159.

Wilson, S., The Ω-spectrum for Brown-Peterson cohomology. Comment. Math. Helv. 48 (1973), 45-55.

Zahler, R., The Adams-Novikov spectral sequence for the spheres. Ann. of Math. (2), 96 (1972), 480-504.

Zabrodsky, A.,

1. Implications in the cohomology of H-spaces. Ill. J. of Math. 14 (1970), 363-375.
2. Secondary operations in the module of indecomposables. Proc. Adv. Study Inst. on Algebraic Topology. Aarhus Denmark (1970), 658-672.
3. Homotopy associativity and finite CW complexes. Topology 9 (1970), 121-128.

4. Cohomology operations and homotopy commutative H-spaces. The Steenrod algebra and its applications. Lecture notes in Mathematics 168, Springer (1970), 308-317.

5. On spherical classes in the cohomology of H-spaces. H-spaces Neuchatel Août 1970, Lecture notes in Math. 196, Springer, 25-33.

6. Secondary operations in the cohomology of H-spaces. Ill. J. of Math. 15 (1971), 648-655.

7. The classification of simply connected H-spaces with three cells. II, Math. Scand. 30 (1972), 211-222.

8. On the construction of new finite CW H-spaces, Inventiones Math. 16 (1972), 260-266.

9. p-equivalences and homotopy type localization in group theory and homotopy theory (Seattle 1974). Lecture notes in Math. 418, Springer.

10. On the homotopy type of prinicpal classical group bundles over spheres. Israel J. of Math. 11 (1972), 315-325.

The following are used as standard notations throughout these notes. We indicate below the page where they are first introduced.

We indicate below the page where the following notions are first introduced in this volume.